Methods in
BIOSTATISTICS

Methods in
BIOSTATISTICS

For
Medical Students and Research Workers

B.K. Mahajan MBBS DPH FCGP FIAPSM
Professor of Preventive & Social Medicine
at Jamnagar (1960–73) and Sevagram (1973–82)
Deputy Director Health Services
erstwhile Mumbai State (1958–60) and
Sr. Consultant, Integrated Child Development Services
Deptt. of Gastroenterology and Human Nutrition Unit
AIIMS, Ansari Nagar, New Delhi-110 029 (1982–87)
Consultant to Delhi Council of Child Welfare
Statistical Check by Sh. B.D. Malhotra & Prof. K.R. Sundram
Head of Deptt. of Biostatistics, AIIMS

Foreword by Dr. V.P. Reddaiah

JAYPEE BROTHERS
MEDICAL PUBLISHERS (P) LTD.
B-3 EMCA House, 23/23B Ansari Road, Daryaganj
Post Box 7193, New Delhi 110 002, India

Branches
Bangalore • Calcutta • Chennai • Mumbai

Published by
Jitendar P Vij
Jaypee Brothers Medical Publishers (P) Ltd
EMCA House, 23/23B Ansari Road, Daryaganj, **New Delhi** 110 002, India
Phones: +91-11-23272143, 23272703, 23282021, 23245672
Fax: +91-11-23276490, +91-11-23245683
e-mail: jaypee@jaypeebrothers.com
Visit our website: www.jaypeebrothers.com

Branches

- 2/B, Akruti Society, Jodhpur Gam Road Satellite
  Ahmedabad 440 009 (MS) Phone: +91-79-30988717
- 202 Batavia Chambers, 8 Kumara Krupa Road
  Kumara Park East, **Bangalore** 560 001
  Phones: +91-80-22285971, +91-80-22382956, +91-80-30614073
  Tele Fax : +91-80-22281761   e-mail: jaypeebc@bgl.vsnl.net.in
- 282 IIIrd Floor, Khaleel Shirazi Estate, Fountain Plaza
  Pantheon Road, **Chennai** 600 008, Phones: +91-44-28262665, 28269897
  Fax: +91-44-28262331 e-mail: jpchen@eth.net
- 4-2-1067/1-3, 1st Floor, Balaji Building, Ramkote Cross Road
  Hyderabad 500 095, Phones: +91-40-55610020, +91-40-24758498
  Fax: +91-40-24758499   e-mail: jpmedpub@rediffmail.com
- 1A Indian Mirror Street, Wellington Square
  Kolkata 700 013, Phones: +91-33-22456075, +91-33-22451926
  Fax: +91-33-22456075   e-mail: jpbcal@cal.vsnl.net.in
- 106 Amit Industrial Estate, 61 Dr SS Rao Road Near MGM Hospital,
  Parel, **Mumbai** 400 012, Phones: +91-22-24124863, 24104532, 30926896
  Fax: +91-22-24160828   e-mail: jpmedpub@bom7.vsnl.net.in
- "KAMALPUSHPA" 38, Reshimbag Opp. Mohota Science College,
  Umred Road **Nagpur** 440 009 (MS)
  Phone: +91-712-3945220, e-mail: jpmednagpur@rediffmail.com

Methods in Biostatistics, Sixth Edition

© 1997, BK Mahajan

All rights reserved. No part of this publication should be reproduced, stored in a retrieval system, or transmitted in any form or by any means: electronic, mechanical, photocopying, recording, or otherwise, without the prior written permission of the author and the publisher.

> This book has been published in good faith that the material provided by author is original. Every effort is made to ensure accuracy of material, but the publisher, printer and author will not be held responsible for any inadvertent error(s). In case of any dispute, all legal matters are to be settled under Delhi jurisdiction only.

First Edition   : 1967
Second Edition : 1970
Third Edition   : 1981
Fourth Edition : 1984
Fifth Edition   : 1989
Sixth Edition   : 1997
  Reprint   : 1999, 2003, 2004, 2005, 2006

ISBN 81-7179-520-X

Typeset at   JPBMP typesetting unit
Printed at   Gopsons Papers Ltd., A-14, Sector 60, Noida

"Those whom the gods love die young"

LALIT

(Born on Tuesday, September 12, 1944)

This work is dedicated to the loving memory of Author's only son, LALIT who lost his life on way to USA for higher studies in mechanical engineering, in the Air India Boeing accident near Mont Blanc on January 24, 1966, along with 116 others including the world-renowned Atomic Energy Scientist, Dr. Homi Bhabha.

About the Author

After a brilliant academic career, Dr. B.K. Mahajan got a wide and rich experience as a Senior Health Administrator for 12 years in old Bombay state and over 22 years as Professor of Preventive and Social Medicine. He was Sr. Consultant, ICDS Aug. 1982 to Dec. 1987 and is Consultant to DCCW now.

In 1961-62 he was deputed to visit medical schools, in the UK and Europe to study teaching of P and SM. He has read several papers at All India Conferences and published many of them in Journals of repute, apart from having written book on his subject. From 1973 to 1982 he served as Professor of P & SM at Sevagram (Wardha) where he was continued as life long Professor Emeritus. Since 1982 to 1988 he was selected as Sr. Consultant for ICDS at AIIMS, New Delhi to serve the entire country.

He was honoured to deliver Dhanwantri Oration at IAPSM Conference, Bangalore in 1976, Late Rama Murthi Guest Oration at Hyderabad in 1980 and Dr. Rameshwar Sharma Oration at Bikaner in 1983. He was Secy. of IAPSM in 1979-80, first President of IAPSM in 1983 and is President of Ind. Asso. of Comm. Diseases from 1980.

He was an Indian Delegate to Seminar on Medical Education in SEA Countries, at Pokhra (Nepal) in 1976.

He has travelled widely in India, UK, Europe, USA and Japan to study the problems of Medical Education in general and of teaching Preventive and Social Medicine/Community Medicine and Biostatistics in particular.

viii Methods in Biostatistics

The very next day after retiring from AIIMS he was requested to join as Medical Consultant to Delhi Council of Child Welfare where he is still working in 1997 and DCCW wishes him to continue as long as health permits.

He has been active member of IMA too and is playing a very vital role in various activities of Delhi Medical Association even now.

Foreword

It is my proud privilege to write the foreword to *Methods in Biostatistics* 6th Edition by Dr. B.K. Mahajan, who is the grand old man of Preventive and Social Medicine. Most of the Biostatistics books available to Indian students are by foreign nonmedical statisticians who use mathematical formulas which make medical people averse towards statistics. They fail to understand and appreciate the use of statistics in their continuing education by literature review, practice and research.

But this book has been written by an eminent medical person who understands the mentality of the medical people towards statistics. He has written this book from the perspective of medical people in simple language with very little mathematical formulas and Indian examples. This makes medical people accept and master statistics. The fact that this book has come to 6th edition in 30 years is an eloquent testimony to the popularity of the book.

This book is a great boon to medical students both undergraduates and postgraduates who want to be self significant in statistical analysis. It is a book that every medical person and medical college library must possess. I wish him success in his endeavour and happy to release this edition.

Dr. V.P. Reddaiah
Prof & Head
Centre for Community
Medicine, AIIMS
New Delhi

Preface to the Sixth Edition

Scientific methods of statistical analysis, which were almost unknown a few years ago, today provide very effective data for understanding and in depth study of any branch of knowledge. It facilitates interpretation of facts and application in medical sciences.

Fruits of any new technique or challenge and different modalities in treatment have to be compared. I am happy to bring the sixth edition in 1997 after the fifth in 1989. After each edition reprints had to be prepared by my publisher Jaypee Brothers in view of ever increasing demand in India and abroad because of its easy and simple understanding. There is no such averseness as existed in the past. All medical research workers too realise application of biostatistics as a science for proper evaluation of the work done-be it thesis or presentation of paper, read or published in journals of repute. There is continuous rising demand for a simple book like this, in which mathematical complexities are avoided and students can easily follow the biostatistics as a science of variation. Simple and familiar examples have been provided abundantly.

Use of computers for which special chapter has been added in brief has greatly facilitated the analysis of the work done in medical field and biosciences too.

I am continuing this job in loving memory of my son since 1967. Practically all chapters needed a revision after such a long interval of eight years specially in demography and vital statistics and I have done it.

B.K. Mahajan

Acknowledgements

I am grateful to Dr Prof Reddaiah for writing Foreword to this edition.

I am very grateful to the experts whose remarks are given at the end of the title of this book. I can not help mentioning my gratefulness to late Shri BD Malhotra, Statistician, Prof KR Sundaram, Head of Biostatistical Department in AIIMS and Shri DA Shah, Statistician in Medical College, Surat.

I highly appreciate the all long cooperation and genuine support given by Shri Jitendar P Vij, Managing Director and his younger brother Shri Pawaninder P Vij, Director of Jaypee Brothers Medical Publishers Pvt. Ltd. who are now perhaps the largest medical publishers in India. They render all help to experts in the field of medicine to publish their works. I would be failing in my duty if I don't express my gratitude to Mr PG Bandhu (Director). It is because of their cooperation that I could complete the present volume.

Dr KK Agarwal, Cardiologist at Mool Chand Hospital has given all help despite his busy schedule. A new chapter on computer is due to his help. Dr (Col) BR Chopra, Cardiology Department gave a lot of encouragement by his ever smiling face.

Dr Bir Singh, Additional Professor of Community Medicine, AIIMS also joined me in updating demography and vital statistics and going through the initial draft but because of his other heavy involvements he did as much as he could.

I am grateful to Dr G Anjneyulu, Prof and Head of PSM, Hyderabad, Maj HS Ratti, MD (PSM), Dy Director (MHS), Dr Nitika Arora and Manu her brother, Shri Praful, Mrs Sapna Bhasin, Shri Jeya Shanker and Km Alpna. I am sincerely thankful to the dedicated team of Jaypee Brothers for their whole hearted cooperation. Dr Sultan Ahmed and Ms Mubeen Bano definitely deserve a pat on their back.

Last but not the least I must sincerely thank Smt Anjna Devi Mahajan my wife and Mrs Nandini Churamani my grand daughter for sparing so much time in this dedicated work.

Contents

	Foreword by Dr. V.P. Reddaiah	*ix*
	Preface to the Sixth Edition	*xi*
1.	**Introduction**	1

Application and uses of Biostatistics as a Science, as figures, Scope, Common statistical terms, Notations

2. **Sources and Presentation of Data** — 10

Qualitative (or discrete) data, Quantitative (or continuous data: Methods of presentation—Tabulation, Frequency distribution drawings, Quantitative data, Qualitative data

3. **Measures of Location-Averages and Percentiles** — 35

Measures of central tendency—averages, Mean, Median, Mode, Measures of location—Percentiles, Graphic method, Arithmetical method, Application and uses of percentiles

4. **Variability and Its Measures** — 58

Types, Biological, Real, Experimental, Measures of variability, Range, Semi-interquartile range (Q), Mean deviation, Standard deviation (SD), Coefficient of variation (CV)

5. **Normal Distribution and Normal Curve** — 77

Demonstration of a normal distribution, Normal curve, Standard normal deviate (Z), Asymmetrical distributions

6. **Sampling** — 88

Representative sample, Precision (sample size), Sample bias, Sampling techniques, Simple random sampling, Systematic, Stratified, Multistage, Cluster, Multiphase

7. **Probability (Chance)** — 103

Addition law of probability, Multiplication law, Binomial probability distribution, Probability chance from shape of normal distribution or normal curve

8. **Sampling Variability and Significance** — 115

Sampling distribution, Significance, Estimation of population parameter, Testing statistical hypotheses, Type I and Type II errors, Tests of significance, Z-test, One-tailed and two tailed tests

9. **Significance of Difference in Means**   130

Standard error of mean, (SEX), Applications and uses, Standard error of difference between two means of large samples, Small samples, t-test, Unpaired, Paired, Variance ratio test, Analysis of variance test

10. **Significance of Difference in Proportions of Large Samples**   156

Standard error of proportion (SEP), Application and uses, Standard error of difference between two proportions, SE (p_1-p_2)

11. **The Chi-square Test**   168

Alternate test to find the significance of difference in two or more than two proportions, As a test of association between two events in binomial or multinomial samples, As a test of goodness of fit, Calculation of χ^2 value, Restrictions in application of χ^2 test, Yates' correction

12. **Correlation and Regression**   186

Calculation of correlation coefficient from ungrouped series, Regression, Calculation of regression coefficient (b), Regression line, Standard deviation of the Y measurements for the regression line

13. **Designing and Methodology of an Experiment or a Study**   204

Steps and methodology, Format for presentation of any research work

14. **Demography and Vital Statistics**   214

Demography—Static demography, Dynamic demography, Collection of demographic data, Population census, Records of vital statistics, Sample registration system (SRS), Survey of causes of death (rural), Definitions of vital events, Medical certification of the cause of death, Collection of vital statistical data above the state level, Shortcomings in registration of vital statistics, Records of health departments, Records of health institutions, Table 14.1 Format for monthly hospital abstracts

15. **Measures of Population and Vital Statistics**   236

Measures of population, Growth of population, Population density, Population distribution by age and sex in India, Measures of vital statistics, Measures of fertility, Fertility rates, Reproduction rate, Measures of marital status, Measures of morbidity, Duration of illness, Measures of mortality, Crude death rate, Standardised death rates,

Specific death rates, Application of rates and ratios of vital indices, Monitoring of family planning programme, Monitoring of MCH services, Monitoring of illness in the community.

Comprehensive indicators that measure health, Targets for,'Health for all by the year 2000 AD'

16. Life Table 275

Uses and application: Construction of a life table, Modified life table

17. Exercises 292

Exercises are given chapterwise.
Some required descriptive answers and other medical calculation. Answers to exercises

18. Computers in Medicine 320

Uses of computer (i) As applied to methods in biostatistics, (ii) Application in community or public health care and management, (iii) Their utility in hospitals, nursing homes and clinics of academic experts, (iv) As a great asset in research in medical practice

Appendices

I.	Table of Unit Normal Distribution	327
II.	Table of 't'	329
III.	Table of Variance Ratio	330
IV.	Table of χ^2	332
V.	Table of Correlation Coefficient	333
VI.	Random Number Tables	334
VII.	International Classification of Disease	336

Bibliography 337

Index 339

Chapter 1

Introduction

Statistic or *datum* means a measured or counted fact or piece of information stated as a figure such as height of one person, birth of a baby, etc.

Statistics or *data* would be plural of the same, stated in more than one figures such as height of 2 persons, birth of 5 babies, etc. They are collected from experiments, records and surveys, in all walks of life such as economics, politics, education, industry, business, administration, etc. Medicine too, including Preventive Medicine and Public Health, is one such field.

Statistics though apparently plural, when used in a singular sense, is a *science of figures*. It is a field of study concerned with techniques or methods of collection of data, classification, summarising, interpretation, drawing inferences, testing of hypotheses, making recommendations, etc. when only a part of data is used.

In any book on statistical methods the word *statistics* is used both as a plural of statistic and as a science of figures.

BIOSTATISTICS

Biostatistics is the term used when tools of statistics are applied to the data that is derived from biological sciences such as medicine.

Any science demands precision for its development, and so does medical science. For precision, facts, observations or measurements have to be expressed in figures.

"When you can measure what you are speaking about and express it in numbers, you know something about it but when

you cannot measure, when you cannot express it in numbers, your knowledge is of meagre and unsatisfactory kind."
... Lord Kelvin

Everything in medicine be it research, diagnosis or treatment, depends on counting or measurement. High or low blood pressure has no meaning, unless it is expressed in figures. Incidence of tuberculosis or death rate in typhoid is stated in figures. Enlargement of spleen is measured in fingers' breadth. Thus medical statistics or biostatistics can be called **quantitative medicine.**

In nature, blood pressure, pulse rate, action of a drug or any other measurement or counting varies not only from person to person but also from group to group. The extent of this variability in an attribute or a character, whether it is by chance, i.e., biological or normal, is learnt by studying statistics as a *science*.

Comparison of a variable in two or more groups is of great importance in applied scientific practice of medicine, e.g., infant mortality rate in developing countries like India was around 73 per thousand live births in 1994 while in developed countries like the USA, UK and Japan, the rates have gone down to about 5 per thousand live births per year due to external factors like socio-economic advancement, better application of scientific knowledge in medicine or improved health services. Rise in pulse rate noted after an injection of a drug may be by chance or due to the effect of drug.

Variation more than natural limits may be pathological, i.e., abnormal due to the play of certain external factors. Hence, biostatistics may also be called a **science of variation**. The data after collection, lying in a haphazard mass are of no use, unless they are properly sorted, presented, compared, analysed and interpreted. They mean something more than figures, give a dimension to the problem and even suggest the solution. For such a study of figures, one has to apply certain mathematical techniques called **statistical methods**, such as calculation of standard deviation, standard error and preparation of a life table. Though these methods are quite simple and general in application, medicos follow them only when they are put in a familiar way giving day-to-day medical examples. Moreover, medical statistics merit special attention as they deal with human beings and not with material objects or lower animals. Medical observer has to give his opinion or make an impression after applying these methods.

"General impressions are never to be trusted. Unfortunately when they are of long standing nature, they become fixed rules of life and assume a prescriptive right not to be questioned. Consequently those who are not accustomed to original enquiry, entertain a hatred and horror of statistics. They cannot endure the idea of submitting their sacred impressions to cold blooded verification. But it is the triumph of scientific men to rise superior to such superstitions, to desire tests, by which the value of their beliefs may be ascertained and to feel sufficiently, masters of themselves to discard contemptuously whatever may be found untrue."

... Francis Galton

A medical student should not depend on a statistician for the statistical analysis. For professional interpretation of his results, he should learn the application of methods himself which do not require knowledge of mathematics higher than what he or she had acquired at school. However, he or she should take the guidance of a qualified statistician right from the beginning of any scientific study till drawing the conclusions. **Medical Statistics** go under different names when applied in different fields such as:

Health statistics in public health or community health.

Medical statistics in medicine related to the study of defect, injury, disease, efficacy of drug, serum and line of treatment, etc.

Vital statistics in demography pertaining to vital events of births, marriages and deaths. These terms are overlapping and not exclusive of each other.

Application and Uses of Biostatistics as a Science

1. *In Physiology and Anatomy*
    i. To define what is normal or healthy in a population and to find *limits of normality* in variables such as weight

and pulse rate—the mean pulse rate is 72 per minute but up to what limits it may be normal on either side of mean has to be established with certain appropriate techniques.

ii. To find the *difference between means and proportions* of normal at two places or in different periods. The mean height of boys in Gujarat is less than the mean height in Punjab. Whether this difference is due to chance or a natural variation or because of some other factors such as better nutrition playing a part, has to be decided.

To find the *correlation* between two variables X and Y such as height and weight—whether weight increases or decreases proportionately with height and if so by how much, has to be found.

2. *In Pharmacology*

i. To find the action of drug—a drug is given to animals or humans to see whether the changes produced are due to the drug or by chance.
ii. To compare the action of two different drugs or two successive dosages of the same drug.
iii. To find the relative potency of a new drug with respect to a standard drug.

3. *In Medicine*

i. To compare the efficacy of a particular drug, operation or line of treatment—for this, the percentage cured, relieved or died in the experiment and control groups, is compared and difference due to chance or otherwise is found by applying statistical techniques.
ii. To find an association between two attributes such as cancer and smoking or filariasis and social class—an appropriate test is applied for this purpose.
iii. To identify signs and symptoms of a disease or syndrome. Cough in typhoid is found by chance and fever is found in almost every case. The proportional incidence of one symptom or another indicates whether it is a characteristic feature of the disease or not.

4. In Community Medicine and Public Health

i. To test usefulness of sera and vaccines in the field—percentage of attacks or deaths among the vaccinated subjects is compared with that among the unvaccinated ones to find whether the difference observed is statistically significant.
ii. In epidemiological studies—the role of causative factors is statistically tested. Deficiency of iodine as an important cause of goitre in a community is confirmed only after comparing the incidence of goitre cases before and after giving iodised salt.

In public health, the measures adopted are evaluated. Lowering of morbidity rate in typhoid after pasteurisation of milk may be attributed to clean supply of milk, if it is statistically proved. Fall in birth rate may be the result of family planning methods adopted under National Family Welfare Programme or due to rise in living standards, increasing awareness and higher age of marriage.

Thus, by learning the methods in biostatistics, a student learns to evaluate articles published in medical journals or papers read in medical conferences. He understands the basic methods of observation in his clinical practice or research.

"He who accepts statistics indiscriminately, will often be duped unnecessarily. But he who distrusts statistics, indiscriminately will often be ignorant, unnecessarily." (WA Wallis and HV Roberts, in Nature of Statistics. The Free Press, New York, 1965).

Application and Uses of Biostatistics as Figures

Health and vital statistics are essential tools in demography, public health, medical practice and community services. Recording of vital events in birth and death registers and diseases in hospitals is like book keeping of the community, describing the incidence or prevalence of diseases, defects or deaths in a defined population. Such events properly recorded form the eyes and ears of a public health or medical administrator, otherwise it would be like sailing in a ship

without compass. Thus, biostatistics as a science of figures will tell:
   a. What are the leading causes of death?
   b. What are the important causes of sickness?
   c. Whether a particular disease is rising or falling in severity and prevalence?
   d. Which age group, sex, social class of people, profession or place is affected the most?
   e. The levels or standards of health reached.
   f. Age and sex composition of population in a community.
   g. Whether a particular population is rising, falling, ageing or ailing?
   h. Which health programme should be given priority and what will be the requirements for the same?

Scope

In this handbook, an attempt is made to highlight the basic principles of statistical methods or techniques for the use of medical students. The approach is to equip medicos and other users to the extent that they may be able to appreciate the utility and usefulness of statistics in medical and other biosciences. Certain essential bits of methos in biostatistics, must be learnt to understand their application in diagnosis, prognosis, prescription and management of diseases in individuals and community. The subject forms an integral part of all disciplines of medicine as explained already.

Numerous examples have been worked out in this text and the various steps of calculations have been elaborated. Use of various statistical parameters like mean, standard deviation, standard error, correlation coefficient, etc. and their application in various statistical tests like Z, χ^2, 't', etc. have been explained. Situations in which various tests should be applied are also given in detail. Still simplicity has been the watchword and intricacies have been avoided, to minimise antipathy or averseness on the part of medicos. All previous editions of this book have been found to be popular among the students of other biosciences as well such as zoology, botany, humanities, agriculture, anthropology, etc.

Common Statistical Terms

One should remember before learning the methods in biostatistics, some terms used, their symbols and notations and refer as and when needed.

1. *Variable* A characteristic that takes on different values in different persons, places or things. A quantity that varies within limits such as height, weight, blood pressure, age, etc. It is denoted as X and notation for orderly series as $X_1, X_2, X_3, ..., X_n$. The suffix n is symbol for number in the series. Σ (sigma) stands for summation of results or observation.

2. *Constant* Quantities that do not vary such as $\pi = 3.141$, $e = 2.718$. They do not require statistical study. In biostatistics, mean, standard deviation, standard error, correlation coefficient and proportion of a particular population are considered as constant.

3. *Observation* An event and its measurements such as blood pressure (event) and 120 mm of Hg (measurement).

4. *Observational unit* The source that gives observations such as object, person, etc. In medical statistics the term individuals or subjects is used more often.

5. *Data* A set of values recorded on one or more observational units.

6. *Population* It is an entire group of people or study elements—persons, things or measurements for which we have an interest at a particular time. Populations are determined by our sphere of interest. It may be *infinite* or *finite*. If a population consists of *fixed* number of values, it is said to be *finite*. If population consists of an *endless* succession of values, the population is an *infinite* one. It has to be fully defined such as all human beings, all families joint or nuclear, all women of 15–45 years of age or only married women, all patients, all doctors in service or in practice and so on. Such a population invariably gives *qualitative data*. If it is finite or limited in number it can easily be counted.

A statistical population may also be birth weights, haemoglobin levels, readings of a thermometer, number of

RBCs in the human body, etc. Such a population mostly gives *quantitative data*. It is *finite* or small in number or *infinite* or unlimited in number that cannot be easily counted.
7. *Sampling unit* Each member of a population.
8. *Sample* It maybe *defined as a part of a population*. It is a group of sampling units that form part of a population, generally selected so as to be representative of the population whose variables are under study. There are many kinds of sample that can be selected from a population. Various methods employed are described later in the book.
9. *Parameter* It is a *summary value* or constant of a variable that describes the *population* such as mean, variance, correlation coefficient, proportion, etc. Familiar examples are mean height, birth rate, morbidity and mortality rates, etc.
10. *Statistic* It is a *summary value* that describes the *sample* such as its mean, standard deviation, standard error, correlation coefficient, proportion, etc. This value is calculated from the sample and is often applied to population but may or may not be a valid estimate of population. Though not desirable, parameter and statistic are often used as synonyms.
11. *Parametric test* It is one in which population constants as described above are used such as mean, variances, etc. and data tend to follow one assumed or established distribution such as normal, binomial, Poisson, etc.
12. *Non-parametric tests* Tests such as χ^2 test, in which no constant of a population is used. Data do not follow any specific distribution and no assumptions are made in nonparametric tests, e.g. to classify good, better and best you allocate arbitrary numbers or marks to each category.

Notations for a Population and Sample Values

Roman letters are used for statistics of samples and Greek for parameters of population.

Common notations are:

Summary Value	Sample statistics	Population parameters
Mean	$\bar{X}$	μ
Standard deviation	s	σ
Variance	s^2	σ^2
Proportion	p	P
Complement of proportion	q	Q

Other symbols commonly used are:

=	:	Equal to
>	:	Greater than
<	:	Lesser than
Z	:	The number of standard deviations from the mean or standard normal deviate/variate
%	:	Per cent
γ	:	Pearson's correlation coefficient
ρ	:	Spearman's correlation coefficient
O	:	Observed number
E	:	Expected number
d.f. or f	:	Degrees of freedom
K	:	Number of groups or classes
P	:	Probability

Chapter **2**

Sources and Presentation of Data

The main *sources* for collection of medical statistics are:
1. Experiments
2. Surveys and
3. Records

Nos. 1 and 2 are specially applied to generate data needed for specific purposes while the records provide ready-made data for routine and continuous information.

1. Experiments

Experiments are performed in the laboratories of physiology, biochemistry, pharmacology and clinical pathology or in the hospital wards for investigations and fundamental research. The data collected with specific objective by one or more workers are compiled and analysed. The results are made use of in preparation of dissertations, theses and scientific papers of publication in scientific journals and books.

2. Surveys

Surveys are carried out for epidemiological studies in the field by trained teams to find the incidence or prevalence of heatlh or disease situations in a community such as incidence of malaria or prevalence of leprosy. They are also made use of in operational research, such as assessment of existing conditions and their place to follow a programme or to study the merits of different methods adopted to control a disease, such as malaria or in eradication programmes, such as smallpox. Thus, they provide useful information on:

a. Changing trends in health status, morbidity, mortality, nutritional status or environmental hazards, health practices, etc.
b. Provide feedback which may be expected to modify policy and system itself and lead to redefinition of objectives.
c. Provide timely warning of public health hazards.

3. Records

Records are maintained as a routine in registers or books over a long period of time for various purposes such as for vital statistics—births, marriages and deaths and for illnesses in hospitals. Data thus collected, are made use of in demography (population studies) and public health practice. The data collected are qualitative, obtained by counting the individuals with same attribute such as still births. persons suffering from peptic ulcers, etc. These events are recorded or registered in medical or public health services, as a routine. Regular administrative machinery is necessary to record or register the different events.

STATISTICAL DATA

The data collected may be for profile or prospective studies at local, state, national or international level. They are analysed to assess changes in health or disease situations in the community or population by standard parameters. The statistical data obtained from the above sources can be divided into two broad categories:
1. Qualitative
2. Quantitative

1. Qualitative (or Discrete) Data

In such data there is no notion of magnitude or size of the *characteristic* or attribute as the same cannot be measured. They are classified by counting the individuals having the same characteristic or *attribute* and not by measurement. There is only one variable, i.e., the number of persons and not the characteristic. Persons with the same characteristic are counted to form specific groups or classes such as attacked, escaped, died, cured, relieved, vaccinated, males, young, old, treated, not treated, on drug, on placebo, etc. The characteristic such as being attacked by a disease or being treated by a drug is not a measurable variable, only the frequency of persons treated or diseased, varies. By one line of treatment 20 survive out of 25 while on the other line, 15 may survive out of 25. The characteristic, i.e., survival, remains the same but the number or frequency of survivors varies. Qualitative data are **discrete** in nature such as number of deaths in different years, population of different towns, persons with different blood groups in a population, and so on.

In medical studies such data are mostly collected in pharmacology to find the action of a drug, in clinical practice to test or compare the efficacy of a drug, vaccine, operation or line of treatment, and in demography to find births, deaths, still births, etc. The results thus obtained are expressed as a ratio, proportion, percentage or a rate. The statistical methods commonly employed in analysis of such data are standard error of proportion and chi-square tests as discussed in later chapters.

2. Quantitative (or Continuous) Data

In statistical language any character, characteristic or quality that varies is called **variable**. In qualitative data as explained above the characteristics such as births, attacked or died do not vary, only the frequency, i.e., the number born or attacked varies. The quantitative data have a *magnitude*. The characteristic is measured either on an interval or on a ratio scale. In such classification, there are two variables—the *characteristics* such as *height* and the *frequency*, i.e., the number of persons with the same characteristic and in the same range. Height varies from person to person, it may be 150 cm in one and 160 cm in another person of the same age and sex. Number of persons with 150 cm, or in the range of 150 to 152 cm, may be 10 while those with height 160 cm or in the range of 160 to 162 cm, may be 20. Thus, we find the *characteristic* as well as the *frequency* both vary from person to person as well as from group to group.

The quantitative data obtained from characteristic variable are also called **continuous** data as each individual has one measurement from a continuous spectrum or range such as body temperature from 35°C to 42°C, height 150 cm to 180 cm, pulse rate from 68 per minute to 84 per minute and so on. The observations ascend or descend from 0 or any starting point in the range of spectrum, such as systolic blood pressure of 100 individuals rising from lowest 90 mm of Hg to the highest 150 mm of Hg.

The characteristic may be measurable in whole numbers and fractions such as chest circumference—33 cm, 34.5 cm, 35.2 cm, 36 cm, 37.3 cm, and so on or it may be measurable or countable in discrete whole numbers only, such as pulse rate; cholesterol; blood pressure; ESR; blood sugar; etc. In medical studies, such statistics are mostly collected in anatomy and physiology, i.e., in health, to define the normal or to find the limits of deviation from the normal in healthy persons. When the measurement or counting crosses the normal limits, it becomes unusual and may indicate pathology. Some of the statistical methods employed in analysis of such data are mean, range, standard deviation, coefficient of variation and correlation coefficient.

Data collected and **compiled** from experimental work, records and surveys should be accurate and complete. They must be checked for accuracy and adequacy before processing further. So far they lie in masses or are scattered in the records, in other words they are mixed and unsorted. Next step, therefore, is sorting or classification of the data into characteristic groups or classes as per age, sex, social class, attacks, etc. **They should** be presented in such a way that data should:
— become concise without losing the details;
— arouse interest in the reader;
— become simple and meaningful to form impressions;
— need few words to explain;
— define the problem and suggest the solution too; and
— become helpful in further analysis.

For good presentation of data, full labelling, simplicity, and honesty are essential requirements.

METHODS OF PRESENTATION

There are two main methods of presenting frequencies of a variable character or a variable.
A. Tabulation
B. Drawing

A. Tabulation

Tabulation are devices for presenting data from a mass of statistical data. Preparation of frequency distribution table is the first rquirement. Table can be simple or complex depending upon measurement of single set of items or multiple sets of items.

Frequency Distribution Table or Frequency Table

In most of the studies, the information is collected in large quantity and the data should be classified and presented in the form of a frequency distribution table as shown in Tables 2.1 and 2.3. This is a very important step in statistical analysis. It groups large number of series or observations of master table and presents the data very concisely, giving all information at a glance. All the frequencies considered together form the *frequency distribution*. The number of persons in each group is called the frequency of that group. It records how frequently a characteristic or an event occurs in persons of the same group.

The frequency distribution table of most biological variables develops a distribution which can be compared with the standard distributions such as **normal, binomial** or **Poisson**. Tabulation of frequencies may be for:
  I. Qualitative data
  II. Quantitative data

I. Qualitative data

In qualitative data, there is no notion of magnitude or size of attribute, hence the presentation of frequency distribution is very simple because the characteristic is not variable but *discrete* (Tables 2.1 to 2.5).

In these tables, each characteristic such as deaths form one whole group or class and is not split into subgroups or subclasses because there is no range of variability, and no class interval. Death means death and attack means attack, no fractions or parts are there.

Frequency Distribution Table 2.1: Colour choice of medical students of women's saree and men's shirt

Sex	Item	Colour choice					Total
		White	Pink	Blue	Green	Yellow	
Boys	Saree	0	60	65	10	10	145
	Shirt	60	15	60	0	10	145
Girls	Saree	10	5	30	20	0	65
	Shirt	45	0	15	5	0	65
Com-	Saree	10	65	95	30	10	210
bined	Shirt	105	15	75	5	10	210

Source: Author's class in 1976

Frequency Table 2.2: Sexwise case fatality rate in untreated cases of typhoid in punjab before introduction of chloramphenicol

Attribute	Men	Women	Total
Attacks	40	30	70
Deaths	12	8	20
Percentage died	30.0	26.7	28.6

Source: Author's information in late 1940s

Frequency Table 2.3: ICDS blocks in operation in different years in India*

1978	33
1981	246
1986	1612
1992	2616
1993	3066
1995	3907

Source: ICDS CTC report 1975 to 1995, P10

Frequency Table 2.4: Tuberculosis in India in different years

	1961	1981	2001 (Estimated)
Total population in millions	439.23	685.18	1052.50
Infections per 100 population	1.75	2.74	4.21
Annual mortality per 100 cases	0.36	0.57	0.88

Source: ADGH (TB) Govt of India, New Delhi

Table 2.5: Showing sexwise distribution of leading sites of cancer in Surat (1980–85)

Code No	Site	No of cases
Males:		
161	Larynx	213 (9.0%)
150	Oesophagus	194 (8.22%)
146	Oropharynx	162 (6.90%)
	Tonsil and soft palate	109 (4.60%)
141	Tongue	83 (3.51%)
149	Pharynx	67 (2.84%)
158	Nasal cavity	59 (2.50%)
162	Lung	43 (1.82%)
186	Testis	27 (1.14%)
144	Floor of mouth	26 (1.10%)
Females:		
174	Female breast	131 (7.42%)
180	Cervix	175 (5.13%)
161	Larynx	110 (4.66%)
146	Oropharynx	90 (3.81%)
	Tonsil and soft palate	77 (3.26%)
171	Carcinoma of connective and soft tissue	53 (2.24%)
183	Ovary	41 (1.73%)
141	Tongue	41 (1.73%)
151	Stomach	34 (1.44%)
162	Lung	31 (1.31%)

Source: J Ind Med Assoc Vol No. 3, March 1995; P 102

II. Quantitative or continuous data

Presentation of quantitative data is generally more cumbersome because the characteristic having a measured magnitude or size as well as the frequency, i.e., the number of persons are both variable. The data of variable characteristics are *continuous* such as height, weight, pulse rate, bleeding time, etc. They have a range from the lowest to the highest. This range is divided into subranges or groups and subrange frequency called, *class frequency* is noted opposite each group.

Frequency Distribution Table 2.6: Presentation of quantitative data of height in markings

Heights of groups in cm	Markings						Frequency of each group
160–161	++++	++++					10
162–163	++++	++++	++++				15
164–165	++++	++++	++++	//			17
166–167	++++	++++	++++	////			19
168–169	++++	++++	++++	++++			20
170–171	++++	++++	++++	++++	++++	/	26
172–173	++++	++++	++++	++++	++++	////	29
174–175	++++	++++	++++	++++	++++	++++	30
176–177	++++	++++	++++	++++	//		22
178–179	++++	++++	//				12
Total							200

The upper limit denotes up to but not including it. Such as height 162 cm is to be counted in next group of 162–164 cm.

The interval from one subrange to the next subrange is known as group or *class interval* and is equal throughout. As an example, the Frequency Table 2.6 for heights of 200 boys varying from 160 to 180 cm with a class interval of 2 cm has been prepared. The number of boys within the heights falling in a particular group is recorded opposite. Groups should have no fractions for easy calculations as depicted in the Table. The group 160–162 will include height 160 cm and above up to 161.99 cm. Group 162–164 will include height 162 cm and above up to 163.99 cm but less than 164.

Last group 178–180 will include height 178 cm and above up to 179.99 or less than 180 cm. The method of formation of groups should be understood in the beginning only.

Age is a continuous variable but the standard accepted groups are made as 15–19, 20–24, 25–29, 30–34, 35–39, 40–44 and 45–49 years in studying characteristics like fertility or acceptance of family planning method in fertile groups of women in the age range of 15–49 years. Age group 15–19 will include women who have completed 15 years of age and all others who have not completed 20 years, i.e., less than 20 years.

To study mortality and morbidity pattern in population, the following age groups are generally recommended:

0 – <1 (Infants up to 1 year)
1 – 4 (Toddlers 1 to 5 years but not completed 5 years)
5 – 14 (School children < 15 years)
15 – 24 (Adults < 25 years)
25 – 34 (Adults < 35 years)
35 – 44 (Adults < 45 years)
45 – 55 (Adults < 55 years)
55 – 65 (Retirement age)
Above 65 (Old age)

When the data in a variable are discrete and measurable in whole numbers such as persons, ESR, blood pressure, cholesterol, income, etc., the data are presented in tabular form (Table 2.7).

Frequency Distribution Table 2.7: Diastolic blood pressure in persons aged 25–34

Blood pressure	No. of males	Percentage	No. of females	Percentage
70–75	6	6.3	8	14.5
76–80	18	19.0	12	21.9
81–85	46	48.4	25	45.5
86–90	17	17.9	8	14.5
91–95	6	6.3	2	3.6
96–100	2	2.1	–	–
Total	95	100.0	55	100.0

Percentage or relative frequency of each group may be calculated and mentioned in another column opposite the relevant group, such as 11.7% opposite group 8–10, 25.2% opposite group 10–12 and so on (Table 2.8) in order to compare the frequencies of different groups.

Rules for Making a Frequency Distribution Table
1. The class or group interval between the groups should not be too broad or too narrow, e.g., grouping of age should not be yearly or 20 yearly, but 5 yearly. In noting heights of adults, interval of 2 cm is kept and not 5 cm. Too large a group will omit the details and too small will defeat the purpose of making the data concise.
2. The number of groups or classes should not be too many or too few but be ordinarily between 6 and 16 depending on the details necessary and the size of sample.
3. The class interval should be same throughout such as 10 mm in observations of the systolic blood pressure from 80–160 mm of Hg, the groups in ascending order will be 80–89, 90–99, ..., 150–159.
4. The headings must be clear such as "height" in inches or in centimetres "age" in years or months, etc. If the data are expressed as rates, mention per cent or per thousand.
5. The rates and proportions, if given the actual number in the group must also be noted, which may be done in parenthesis.
6. Groups should be tabulated in ascending or descending order, from the lowest value in the range to the highest such as pulse rate 61–65, 66–70, ..., 106–110.
7. If certain data are omitted or excluded deliberately, the reasons for the same should be given.

B. Frequency Distribution Drawings

After classwise or groupwise tabulation, the frequencies of a characteristic can be presented by two kinds of drawings—**graphs** and **diagrams.** They may be shown either by lines and dots or by figures. The drawings are meant for the nonstatistical minded people who want to study the relative values of frequencies of persons or events. For the statistical-minded persons, they are for quick eye reading.

Presentation of quantitative, continuous or measured data is through graphs. The common graphs in use are:
1. Histogram
2. Frequency polygon
3. Frequency curve
4. Line chart or graph
5. Cumulative frequency diagram
6. Scatter or dot diagram.

Presentation of qualitative, discrete or counted data is through diagrams. The common diagrams in use are:
1. Bar diagram
2. Pie or sector diagram
3. Pictogram or picture diagram
4. Map diagram or spot map.

Presentation or Illustration of Quantitative Data

1. *Histogram*

It is a graphical presentation of frequency distribution. Variable characters of the different groups are indicated on the horizontal line (x-axis) called *abscissa* while frequency, i.e., number of observations is marked on the vertical line (y-axis) called *ordinate*. Frequency of each group will form a column or rectangle. Such a diagram is called '*histogram*'

Frequency Distribution Table 2.8: Tuberculin reaction measured in 206 persons

Reaction mm	Frequency	Percentage
8–10	24	11.7
10–12	52	25.2
12–14	42	20.4
14–16	48	23.3
16–18	12	5.8
18–20	8	3.9
20–22	14	6.8
22–24	6	2.9
Total	206	100.0

Sources and Presentation of Data

and is made use of in presenting any quantitative data such as the one given in Table 2.8, obtained after Mantoux Test.

The upper limits denote up to but not including, e.g., observation 10 mm in the first group is to be included in the next group 10–12.

Histogram is an area diagram Area of the rectangles varies with the frequency. The height of rectangle alone will indicate the frequency if the class interval is uniform as in Figure 2.1.

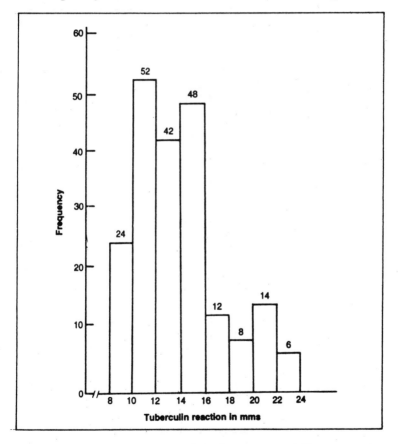

Fig. 2.1: Histogram showing tuberculin reaction in 206 persons, never vaccinated

22 *Methods in Biostatistics*

If the class intervals are different in certain groups then area of the rectangle alone indicates the frequency, e.g. the frequency of the persons in group with size of Mantoux reaction from 16 mm to 24 mm may be presented as one rectangle only (Fig. 2.2). In this case to plot the frequency for this group, divide the total frequency by 4 (40 ÷ 4). Now the horizontal line will start opposite 10 on the vertical line. It would have given an erroneous idea if the horizontal line would start opposite 40 on the vertical line, instead of 10.

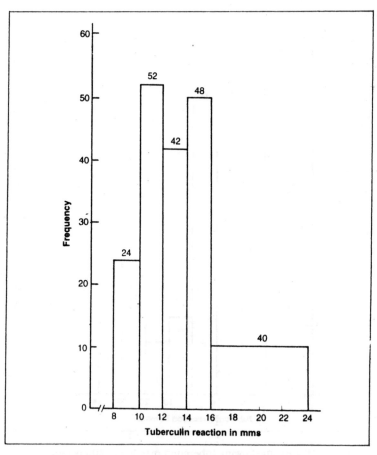

Fig. 2.2: Histogram showing tuberculin reaction in Fig. 2.1 when class interval is not uniform

2. Frequency Polygon

It is again an area diagram of frequency distribution developed over a histogram. Join the mid-points of class intervals at the height of frequencies by straight lines. It gives a polygon, i.e., a figure with many angles (Table 2.8, Fig. 2.3).

In this diagram also, the last frequencies in percentages may be grouped into one and the average of 4 groups in percentage may be plotted. The end of the polygon will now be a straight line parallel to the baseline extending from 16 mm to 24 mm corresponding to the average frequency of this group on the vertical line.

It is used when sets of data are to be illustrated on the same diagram such as birth and death rates, birth of diabetics and non-diabetics, etc.

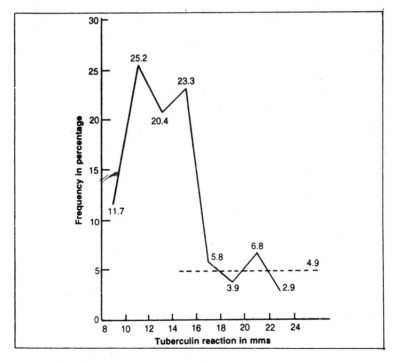

Fig. 2.3: Frequency polygon showing tuberculin reaction in 206 persons never vaccinated

3. Frequency Curve

When the number of observations is very large and group interval is reduced, the frequency polygon tends to lose its angulation giving place to a smooth curve known as *frequency curve*. This provides continuous graph giving the relative frequency for each value of an attribute such as that of height in Figure 2.4 of a normal curve. Such a curve is obtained in normal distribution of individuals in a large sample or of means in population or of differences in pairs of sample means.

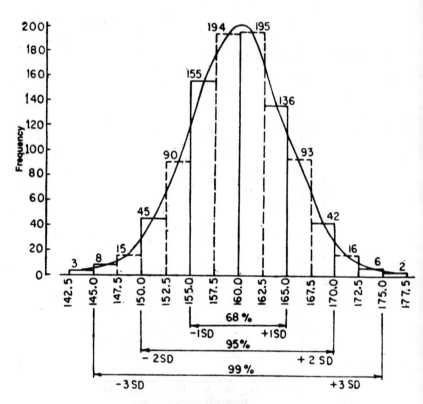

Fig. 2.4: Histogram of 1000 heights in centimetres with normal curve superimposed

4. Line Chart or Graph

This is a frequency polygon presenting variations by line. It shows the trend of an event occurring over a period of time rising, falling or showing fluctuations such as of cancer deaths, infant mortality rate, birth rate, death rate, etc., say from year 1900 to 1960. The class interval may be a month, a year, 5 years or 10 years. Deteriorating or improving trend before and after a public health measure, such as fall of malaria cases each year after DDT spray can be seen at a glance from such a figure. Vertical axis may not start from zero but at some point above when frequencies start at high level as shown in Figure 2.5 indicating the population trend in India. The shape of line chart may alter with change of

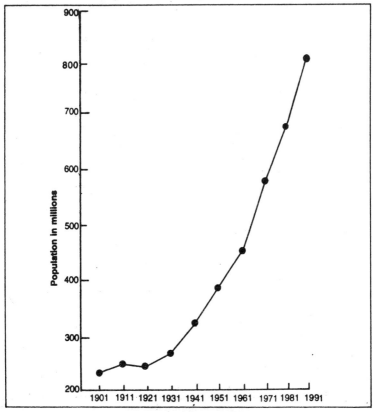

Fig. 2.5: Line chart showing the population trend in India

scale on the vertical or horizontal axis but the trend indicated, remains the same. The proportional change may be there when scales are changed. Figures 2.6 and 2.7

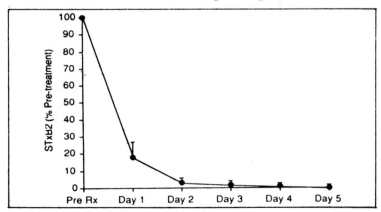

Fig. 2.6: Time to achieve maximum antiplatelet effect with 75 mg aspirin/day, it is maximum in first day then declines unless taken next day again

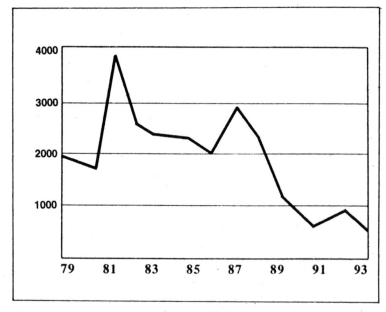

Fig. 2.7: Decrease in poliomyelitis cases in India from 1979 to 1993 after pulse polio immunisation

depicts antiplatelet effects of aspirin and incidence of polio in population after pulse polio immunization.

It is not a good diagram when 0 is suppressed especially in small frequencies. The line is extrapolated as in Figure 2.5.

5. Cumulative Frequency Diagram or 'Ogive'

Ogive is a graph of the cumulative relative frequency distribution. To draw this, an ordinary frequency distribution table in a quantitative data has to be converted into a relative cumulative frequency table. Cumulative frequency is the total number of persons in each particular range from lowest value of the characteristic up to and including any higher group value. It is obtained by cumulating the frequency of previous classes including the class in question (Table 2.9).

The reader is familiar with the cumulative cricket score recorded in a cricket test match after each player, simultaneously with the serial record of the number of runs scored by each individual player.

Table 2.9: Cumulative frequency

Height of groups in cm	Frequency of each group	Cumulative class frequency
160–162	10	10
162–164	15	25
164–166	17	42
166–168	19	61
168–170	20	81
170–172	26	107
172–174	29	136
174–176	30	166
176–178	22	188
178–180	12	200
Total	200	

The cumulative frequencies are plotted corresponding to the group limits of the characteristic. On joining the points by a smooth free hand curve, the diagram made is called **ogive** (Fig. 2.8).

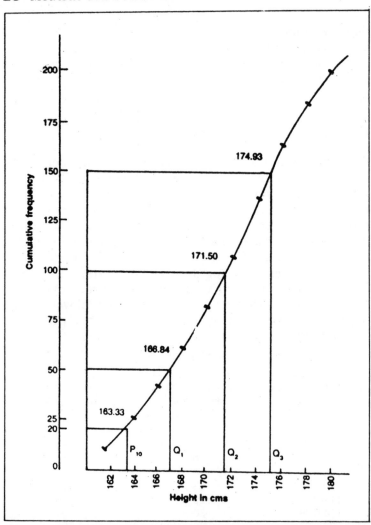

Fig. 2.8: Cumulative frequency diagram showing height values of median (Q_2), first or lower quartile (Q_1), third or upper quartile (Q_3) and tenth percentile (P_{10})

Now one can locate any percentile that divides the series into two parts, e.g., first decile divides the total frequency into 10% and 90%. The calculation and application of percentiles will be discussed in Chapter 3.

6. Scatter or Dot Diagram

It is prepared after tabulation in which frequencies of at least two variables have been cross classified. It is a graphic presentation, made to show the nature of correlation between two variable characters X and Y in the same person(s) or group(s) such as height and weight in men aged 20 years, hence it is also called *correlation diagram*. The characters are read on the base (height) and vertical (weight) axes and the perpendiculars drawn from these readings meet to give one scatter point. Varying frequencies of the characters give a number of such points or dots that show a scatter. A line is drawn to show the nature of correlation at a glance as in Figures 2.9 and 2.10.

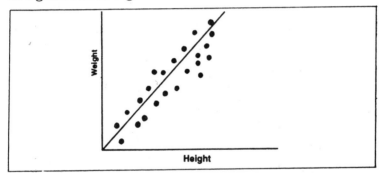

Fig. 2.9: Scatter diagram showing positive correlation

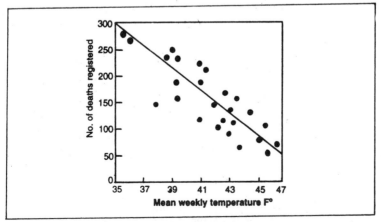

Fig. 2.10: Scatter diagram showing negative correlation

Presentation or Illustration of Qualitative Data

1. Bar Diagram

Length of the bars, drawn vertical or horizontal, indicates the frequency of a character. Bar chart or diagram is a popular and easy method adopted for visual comparison of the magnitude of different frequencies in discrete data, such as of morbidity, mortality, immunisation status of population in different ages, sexes, professions or places. Bars may be drawn in ascending or descending order of magnitude or in the serial order of events. Spacing between any two bars should be nearly equal to half of the width of the bar.

There are three types of bar diagrams: Simple, multiple and proportional bar diagram for comparison of data.

i. **Simple bar diagram** (Figs 2.11 and 2.12)

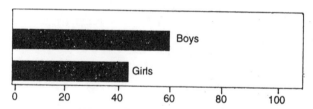

Fig. 2.11: Simple bar diagram showing weight in kg among boys and girls

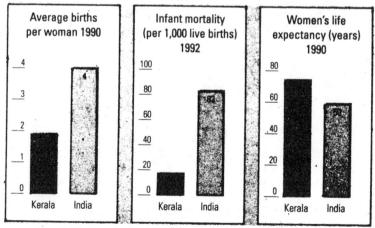

Fig. 2.12: Simple bar diagram showing average birth infant mortality and women's life expectancy

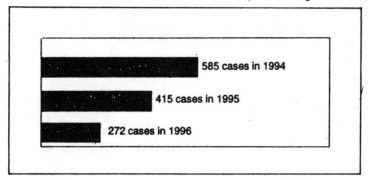

Fig. 2.12b: Fall in reported poliomylitis cases after Pulse Polio Immunisation in Delhi state

ii. **Multiple bar diagram** (Fig. 2.13)

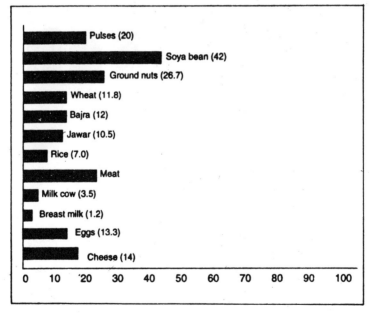

Fig. 2.13: Multiple bar diagram showing protein content of common foods in gm per 100 gm of edible portion (Source: Dr. L.C. Gupta, Food and facts about nutrition)

iii. Proportional bar diagram (Fig. 2.14)

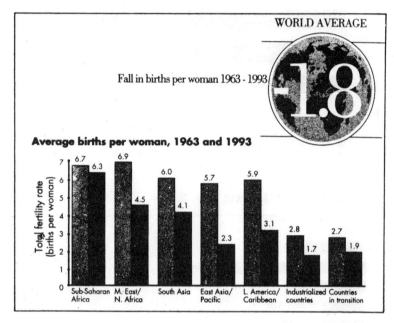

Fig. 2.14: Proportional bar diagram

2. Pie or Sector Diagram

This is another way of presenting discrete data of qualitative characters such as blood groups, Rh groups, age groups, sex groups, causes of mortality or social groups in a population. The frequencies of the groups are shown in a circle (Fig. 2.15). Degrees of angle denote the frequency and area of the sector. It gives comparative difference at a glance. Size of each angle is calculated by multiplying the

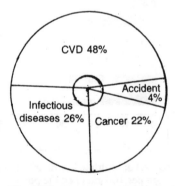

Fig. 2.15: Pie or sector diagram of world death rate due to various diseases (Sources: Dr. KK Agarwal in Medinews, WHO)

class percentage with 3.6, i.e.,

$\frac{360}{100}$ or by the formula $\frac{\text{Class frequency}}{\text{Total observations}} \times 360°$,

Accordingly, the size of angle for blood groups A, B, AB and O gives frequency of the groups such as 26.5%, 34.5%, 17.4% and 31.6%, (total 100), respectively can be drawn as in Figure 2.15, in which pie diagram percentage mortality in the world is indicated by 4 major causes of death.

3. Pictogram or Picture Diagram

It is a popular method to impress the frequency of the occurrence of events to common man such as attacks, deaths, number operated, admitted, discharged, accidents, etc. in a population

Figure 2.16 shows AIDS in developing and indus- trially developed countries. The burden of disease caused by HIV infection is clear.

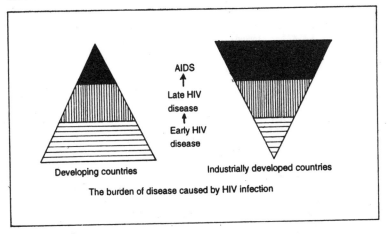

Fig. 2.16: Pictogram of HIV infection and AIDS is clear

4: Map Diagram or Spot Map

These maps are prepared to show geographical distribution of frequencies of characteristic. The figure below in the

34 *Methods in Biostatistics*

statewise map of India indicates the IMR in that state which is lowest 23 in Kerala and highest 126 in Orissa (Fig. 2.17).

Fig. 2.17: Estimated infant mortality rates—1989

Chapter 3

Measures of Location—Averages and Percentiles

Having learnt the methods of collection and presentation of data, we have to understand and grasp the application of mathematical techniques involved in analysis and interpretation of data. Certain symbols and formulae are invariably used in statistical calculations. As medicos, we should learn to apply the formulae straight to our problems without worrying how they have been deduced. Application of methods for analysis is quite easy and we should become familiar with them so as to verify our preconceived ideas or to remove doubts which might arise at the first look of figures collected.

"If a man will begin with certainties, he shall end in doubts': but if he will be content to begin with doubts. he shall end in certainties." ...Francis Bacon

We may first learn methods of analysing quantitative data in which characteristics and frequency are both variable such as calculation of averages, percentiles, standard deviation, standard error, correlation and regression coefficients, etc. They are mostly used in anatomy and physiology to assess what is normal in health as in the case of height. weight. blood pressure, cholesterol in blood, pulse rate, RBC count, etc.

Normal is not the mean or central value but the accepted range of variation on either side of the average or mean, e.g., normal blood pressure is not the mean but mean 120 ± 20.

i.e., a range between 100 and 140. Chances of even higher and lower observations being normal are there.

MEASURES OF CENTRAL TENDENCY—AVERAGES

Information from a series of observations presented by rank in a sample or frequency distribution table is summarised by an observer to get answers to the following three questions:
1. What is its average or central value?
2. How are the other values dispersed around this value or what is the degree of scatter?
3. What is the shape of the distribution, is it normal?

Average value of a characteristic is the one central value around which all other observations are dispersed. In any large series, nearly 50% observations lie above while the remaining 50% lie below the central value. It indicates how the values lie near the centre. In other words, it is a measure of central tendency or concentration of all other observations around the central value. Thus an average value helps:

Firstly, to find most of the normal observations lie close to the central value, while few of the too large or too small lie far away at both ends.

Secondly, to find which group is better off by comparing the average of one group with that of the other, e.g., one finds the average incubation period of cholera is smaller than that of typhoid; income of pleaders is higher than that of doctors; average daily attendance of one hospital is higher than that of another; and so on. After finding the difference, one may reason out why in one group it is more than that in the other.

Average is a general term which describes the centre of a series. There are three common types of averages or measures of central position or central tendency—**mean, median** and **mode**. They are summary indices describing the 'central point or the most characteristic value' of a set of measurements.

Mean

This measure implies arithmetic average or arithmetic mean which is obtained by summing up all the observations and dividing the total by the number of observations.

Example
Erythrocyte sedimentation rates (ESRs) of 7 subjects are 7, 5, 3, 4, 6, 4, 5. Calculate the mean.

$$\text{Mean} = \frac{7+5+3+4+6+4+5}{7} = \frac{34}{7} = 4.86$$

It is the one central value, most commonly used in statistical methods.

Median

When all the observations of a variable are arranged in either ascending or descending order, the middle observation is known as median. It implies the midvalue of series. ESRs of seven subjects are arranged in ascending order—3, 4, 4, (5), 5, 6, 7. The 4th observation (5) is the median in this series. Thus in this example, median is almost equal to the mean. Consider another example of 7 observations in absenteeism of school children in the series 4, 6, 8, (10), 12, 14, 32.

Mean 86/7 = 12.3
Median value = 10

In this case, mean value gives a distorted result as one observation 32 is too large, so the mean as a measure of central tendency should not be considered appropriate. To have a better idea of average, one should ignore unduly high observations such as 32 in the above example. Mean of the remaining observations will be 54/6 = 9.0 which is much closer to the median, i.e. 10 than the mean 12.3 calculated with seven observations.

Median, therefore, is a better indicator of central value when one or more of the lowest or the highest observations are wide apart or not so evenly distributed, e.g.

	Year	
	1974	1977
Mean age at marriage	16.40	15.70
Median age at marriage	15.70	16.00

As per mean, the age at marriage in 1977, has decreased while as per median it has increased which is nearer the truth.

Another good example is the duration of stay in a hospital in general or in a specific disease ward, or disease. In such cases median is a better indicator because the stay may be unduly long in some cases.

Mode

This is the most frequently occurring observation in a series, i.e., the most common or most fashionable, such as 8 mm in tuberculin test of 10 boys given below:

3, 5, 7, 7, 8, 8, 10, 11, 12.

Mode is rarely used in medical studies. Out of the three measures of central tendency mean is better and utilised more often because it uses all the observations in the data and is further used in the tests of significance.

Calculation of Mean

A series of observations is indicated by the letter X and individual observations by $X_1, X_2, ..., X_n$. The mean of series is denoted by $\overline{X}$ (X bar); the number of observations by n and the sum of observations by Σ (sigma) which means sum-up or add-up of all the results.

Formula for calculation of mean is:

$$\text{Mean} = \frac{\text{Total or sum of the observations}}{\text{Number of observations}}$$

$$\overline{X} = \frac{(X_1 + X_2, ..., + X_n)}{n} = \frac{\Sigma X}{n}$$

The mean is calculated by different methods in two types of series, ungrouped and grouped.

Ungrouped Series

In such series the number of observations is small and there are two methods for calculating the mean. The choice depends upon the size of observations in the series.

i. When the observations are small in size, simply add them up and divide by the number of observations.

Example

Tuberculin test reaction of 10 boys is arranged in ascending order being measured in millimetres. Find the mean size of reaction.

3, 5, 7, 7, 8, 8, 9, 10, 11, 12

$$\text{Mean or } \overline{X} = \frac{\Sigma X}{n}$$

$$\text{Mean} = \frac{3 + 5 + 7 + 7 + 8 + 8 + 9 + 10 + 11 + 12}{10}$$

$$= \frac{80}{10} = 8 \text{ mm}$$

ii. When individual observations are large in size, either use a calculator for addition and proceed as in (i) above or assume an arbitrary mean or working origin (w). Find out the difference between each measurement and the assumed mean or working origin, sum up the differences and divide the total by 'n', i.e., the number of observations. This gives the mean of differences or deviations denoted by $\bar{x}$ (small x bar).

$$\text{Now } \bar{x} = \frac{(X_1 - w + X_2 - w, ..., + X_n - w)}{n}$$

$$= \frac{\Sigma (X - w)}{n} = \frac{\Sigma x}{n}$$

Here, X stands for the original series of observations and x for the differences between X and the working origin or assumed mean. Mean of the original observations or $\overline{X} = w + \bar{x}$.

Examples

1. Find the mean incubation period of 9 polio cases given below.

In this case 20 may be taken as the working origin or working mean and 0 may be put opposite that.

X	X-w	x
23	23–20	+3
22	22–20	+2
20	20–20	0
24	24–20	+4
16	16–20	–4
17	17–20	–3
18	18–20	–2
19	19–20	–1
21	21–20	+1
180		+10–10=0

a. By direct method
Number of observations n = 9

$$\overline{X} = \frac{\Sigma X}{n} = \frac{180}{9} = 20$$

b. By assumed mean (w) method

$$\overline{x} = \frac{\Sigma(X-w)}{n} = \frac{10-10}{9} = 0$$

so $\overline{X} = w + \overline{x} = 20 + 0 = 20$

2. Heights in centimetres for 7 school children are given below. Find the mean. 140 may be taken as the working origin to calculate the mean height.

X	X-w	x
148	148–140	8
143	143–140	3
160	160–140	20
152	152–140	12
157	157–140	17
150	150–140	10
155	155–140	15
1065		85

a. By direct method
Number of observations n = 7

$$\overline{X} = \frac{\Sigma X}{n} = \frac{1065}{7} = 152.1$$

b. By assumed mean (w) method

$$\overline{x} = \frac{\Sigma(X-w)}{n} = \frac{85}{7} = 12.1$$

$$\overline{X} = w + \overline{x} = 12.1 + 140 = 152.1$$

If w = 150, calculations will be still easier.

Then, $\overline{x} = \dfrac{\Sigma(X-150)}{n}$

$$\overline{x} = \frac{-2-7+10+2+7+0+5}{7} = \frac{15}{7} = 2.1$$

$$\overline{X} = w + \overline{x} = 150 + 2.1 = 152.1$$

3. The heights of 7 boys aged 14 years are given in cm with fractions in the table that follows. Find the mean.

Take 0 for any convenient working origin or assumed mean such as 150 in the example below:

X Height	X-w = x	Sum of x
145.8	−4.2	
146.9	−3.1	−8.5
148.8	−1.2	
150.0(w)	0	
152.1	+2.1	
153.6	+3.6	+12.7
157.0	+7.0	

$$\bar{x} = \frac{\Sigma(X-w)}{n}$$

$$= \frac{12.7 - 8.5}{7} = \frac{4.2}{7} = 0.6$$

$$\bar{X} = w + \bar{x} = 150 + 0.6 = 150.6 \text{ cm}$$

Grouped Series

When the number of observations is large, the data are arranged in groups and frequency distribution table is prepared first. In all grouped series only *weighted mean* has to be found and not the ordinary mean. Make convenient groups as per the characteristic values and prepare the frequency distribution table. Find the value or weight, contributed by each group separately and multiply the mid value of group with its frequency. Total these product values and then divide by the total number of observations in the sample. This mean is called *weighted mean* or *grand mean* or *mean of means*.

Examples

1. The average income of 10 lady doctors is Rs 400 per month and that of 20 male doctors is Rs 600 per month, calculate the weighted mean or average income of all doctors.

One may consider the average income of all doctors as Rs 500/- if he adds the averages of Rs 400 and Rs 600 and then divides by 2. It is, however, advisable to compute after taking into account the incomes or weights contributed by each group of doctors as shown below:

Total income of 10 lady doctors = mean income of the group × frequency = $\bar{X}_1 \times f_1$ = 400 × 10 = Rs 4000.

Total income of 20 male doctors = $\overline{X}_2 \times f_2 = 600 \times 20$ = Rs. 12000

Total income of all the 30 doctors = 4000 + 12000 = Rs 16000 (sum of weights, i.e. $\Sigma f \overline{X}$)

The weighted mean income of all doctors = $\dfrac{\Sigma f \overline{X}}{n}$

$= \dfrac{16000}{30}$ = Rs 533.3 and not Rs 500/-

Weighted mean is computed in the same way when an event in a qualitative data is expressed in percentages.

2. Calculate overall fatality rate in smallpox from the age wise fatality rate given below:

Age group in years	No. of smallpox cases	Fatality rate per cent
0–1	150	35.33
2–4	304	21.38
5–9	421	16.86
Above 9	170	14.17

It will be wrong to compute the overall rate by adding the rates and dividing by 4.

(35.33 + 21.38 + 16.86 + 14.17) ÷ 4 = 21.94

Instead, we will have to calculate the total deaths in each age group from the fatality per cent given above.

No. of deaths
$= \dfrac{35.33}{100} \times 150 + \dfrac{21.38}{100} \times 304 + \dfrac{16.86}{100} \times 421$
$+ \dfrac{14.17}{100} \times 170 = 213.06$

213.06 deaths have occurred out of 1045 smallpox cases, hence overall fatality rate per cent

$= \dfrac{213.06}{1045} \times 100 = 20.39\%$

Calculation can be simplified by directly multiplying the frequency of smallpox cases with the death rate in each age group and then dividing the total deaths by total smallpox

Measures of Location Averages and Percentiles **43**

cases. This eliminates division as well as multiplication by 100.

$$\frac{150 \times 35.33 + 304 \times 21.38 + 421 \times 16.86 + 170 \times 14.17}{1045}$$

= 20.39

Weighted mean is easy when groups are fewer, size of observations is small and they are not in fractions in a frequency distribution table. No working origin is necessary, one has to simply find the value of each group, i.e., weight, add the group weights and divide by the total number in the sample.

3. Find mean days of confinement after delivery in the following series:

Days of confinement X	No. of patients f	Total days of each group (weights), f X
6	5	30
7	4	28
8	4	32
9	3	27
10	2	20
	18	137

Apply the formula for the weighted mean $\frac{\Sigma f X}{n}$ where X denotes the series of observations, i.e., days of confinement and f is the frequency of patients in each group and n is the total frequency, i.e., the number of patients confined.

Mean days of confinement, therefore

$$= \frac{\Sigma f X}{n} = 137 \div 18 = 7.61$$

Class interval in the days of confinement is taken as 1.
4. Calculate mean when class interval is again one but group characteristics and frequencies are in fractions and

44 Methods in Biostatistics

large in size as in the weight of large number of boys in the table that follows:

Weight of children in kg X	Midpoint or average wt of each group X_g	No. of children in each group f	Weight contributed by each group $f X_g$
60 - < 61	60.5	10	10 × 60.5
61 –	61.5	20	20 × 61.5
62 –	62.5	45	45 × 62.5
63 –	63.5	50	50 × 63.5
64 –	64.5	60	60 × 64.5
65 –	65.5	40	40 × 65.5
66 - < 67	66.5	15	15 × 66.5
Total		240	15310.0

Finding the average weight of each group from the original records may be laborious and they may not be available. The groupwise frequency distribution table is prepared as in example 3. To find the average or mean of each group the mid value of the group is taken as its mean such as 60.5 kg for weight group 60 – < 61. Now the weight contributed by each group (X_g) is found by multiplying with the frequency (f). Apply the formula as in example 3.

$$\frac{\Sigma f X_g}{n} = \frac{15310.0}{240} = 63.79$$

This method of calculation is still laborious and can be further simplified by assuming an arbitrary mean or working origin as was done in the ungrouped series when size of observations was large. This working origin will be the mid value of any group which is taken as 0.

The group values above and below this assumed mean (0) may be reduced to working units of 1, 2, 3, ..., etc., which are multiples of class or group interval in the frequency distribution table.

Measures of Location Averages and Percentiles

Now proceed stepwise as follows:

a. Place 0 opposite any convenient group which may be taken as the working origin (w). It should be the mid value of that group.

b. Find the weight contributed by the group by multiplying frequency (f) with weight in working units ($\bar{x}$), i.e., f x where $x = X_g - w$.

c. Sum up the group weights and divide by the number of persons to calculate mean in working units $x = \dfrac{\Sigma fx}{n}$

d. **Multiply** this mean ($\bar{x}$) by the class interval, 1, 5 or 10 as the case may be; $1 \times \bar{x}$ or $5 \times \bar{x}$ or $10 \times \bar{x}$.

e. **Add** this to the working origin. This gives the mean ($\bar{X}$) in real units $\bar{X} = \bar{x} + w$.

Now mean weight may be found as below:

Weight in kg X	Mid value X_g	Frequency or No. f	Working units x	Group weight fx	Sum of fx $\Sigma(fx)$
60–61	60.5	10	−2	10 × −2	
61–62	61.5	20	−1	20 × −1	
62–63	62.5(w)	45	0	45 × 0	
		75			− 40
63–64	63.5	50	+1	50 × 1	
64–65	64.5	60	+2	60 × 2	
65–66	65.5	40	+3	40 × 3	
66–67	66.5	15	+4	15 × 4	
		165			+350
Total		240			+310

Mean in working units, $\bar{x} = \dfrac{310}{240} = 1.29$

Mean in real units, $\bar{X} = w + \bar{x} = 62.50 + 1.29 = 63.79$

Thus, mean in real units = central value (62.5) in real units of the group against which 0 is placed, + $\bar{x}$ × class interval (1.29 × 1).

46 Methods in Biostatistics

If the class interval is 5 as in the next example, multiply the mean in working units by 5 and add it to the central value of the group or arbitrary working origin (w).

5. Find the mean age at death in years in the following series:

Age of persons at death X	Age in working units x	No. died f	Groupwise year lived f_x	Sum of f_x $\Sigma(f_x)$
0– < 5	–6	100	–600	
5–	–5	40	–200	
10–	–4	20	–80	
15–	–3	25	–75	
20–	–2	20	–40	
25–	–1	25	–25	
30– w = 32.5	0	30	0	–1020
35–	+1	15	+15	
40–	+2	18	+36	
45–	+3	16	+48	
50–	+4	20	+80	
55–	+5	15	+75	
60–	+6	5	+30	
65– < 70	+7	1	+7	+291
Total		350		–729

Mean age at death in working units. $\bar{x} = 5 \times \dfrac{-729}{350}$

= 5 × –2.08 = –10.40 (5 is the group interval with which mean in working units has to be multiplied).

Thus real mean age, $\bar{X}$ = 32.5 (mid value opposite which 0 is placed) –10.4 = 22.1 years.

6. Find the mean weight of 470 infants born in a hospital, in one year, from the following table:

Weight of infants in kg X	Mid value	Weight in working units x	No. of infants f	Group weight fx	Sum of fx Σfx
2.0–2.4	2.2	−2	17	−34	
2.5–2.9	2.7	−1	97	−97	−131
3.0–3.4	w=3.2	0	187	0	
3.5–3.9	3.7	+1	135	−135	
4.0–4.4	4.2	+2	28	+56	+209
4.5 +	4.7	+3	6	+18	
Total			470		+78

Mean weight $\overline{X} = w + \overline{x} \times$ group interval where $\overline{x}$ is the mean in working units.

$$\overline{X} = 3.2 + \frac{78}{470} \times 0.5 = 3.20 + 0.083 = 3.283$$

Essential points to be noted in calculation of mean:
1. Ordinary mean is the arithmetic mean of ungrouped series.
2. To simplify calculations, put 0 against any convenient individual observation in ungrouped series, taking it as an arbitrary working origin or assumed mean.
3. In grouped series put 0 opposite any convenient group, preferably central group and find the mean in working units.
4. Keep in view the mid-value of the group opposite which 0 is placed, while finding the mean in real units.
5. Do not forget to multiply the mean in working units with class interval when it is more than 1.
6. Weighted mean is the average of grouped series in a frequency distribution table.

Calculation of Median

Ungrouped Series

Arrange the observations in the series in ascending or descending order of magnitude. The central observation of the arranged series gives the median.

Formula, $\frac{n+1}{2}$ gives the serial number of median, irrespective of the number (n) of observations in the series, odd or even.

In odd number of observations, e.g. 39, the serial number of median will be $\frac{39+1}{2} = 20$, i.e., the 20th observation is median. In case of even number of observations such as 40, apply the same formula $\frac{n+1}{2} = \frac{40+1}{2} = 20.5$.

Median will be the mean of 20th and 21st observations, e.g. if in 40 patients, the period of stay of 20th patient was 10 days and that of 21st patient was 12 days, the median value of 40 patients as regards period of stay will be $\frac{10+12}{2} = 11$ days.

Grouped Series

To find the median in grouped series simply divide the total observation by 2. If total is 200, 100th observation is median. Even if the total is 201, the 100th observation may be taken as median.

Now, find the value of median observation, such as height of 100th boy out of the total 200 (Table 2.9), by graphic and arithmetical methods. It comes to 171.50 cm and 171.46 cm, respectively, as explained later.

MEASURES OF LOCATION—PERCENTILES

Percentiles Averages discussed so far are measures of central value, therefore, they locate the centre or mid point of a distribution. It may also be of interest to locate other points in the range. Percentiles do that. They are values of a variable such as height, weight, age, etc., which divide the total observations by an imaginary line into two parts, expressed in percentages such as 10% and 90% or 25% and 75%, etc. In all, there are 99 percentiles. Centiles or percentiles are values in a series of observations arranged in ascending order of magnitude which divide the distribution into 100 equal parts. Thus, the median is 50th centile. The

Measures of Location Averages and Percentiles

50th percentile will have 50% observations on either side. Accordingly, 10th percentile should have 10% observations to the left and 90% to the right. But for population in India it is not so. If children at age 3½ years form 10th percentile, it means 10% of entire population is below 3½ years of age and 90% is above that age. Age 20 may be median or 50th percentile and divide the people into one half below 20 years and the other half above 20 years of age. In developed countries 10th percentile may be 7½ years, therefore 10% population will be below 7½ years while in India 7½ years may be 20th or 25th percentile and cover 20% or 25% of the total population. Thus, percentiles are used to divide a distribution into convenient groups. Those in common use are described below:

Quartiles They are three different points located on the entire range of a variable such as height—Q_1, Q_2 and Q_3. Q_1 or lower quartile will have 25% observations of heights falling on its left and 75% on its right; Q_2 or median will have 50% observations on either side and Q_3 or upper will have 75% observations on its left and 25% on its right.

Quintiles Quintiles, four in number divide the distribution into 5 equal parts. So 20th percentile or first quintile will have 20% observations falling to its left and 80% to its right.

Deciles Nine in number divide the distribution into 10 equal parts, first decile or 10th percentile will divide the distribution into 10% and 90% while 9th decile will divide into 90% and 10% and 5th decile will be same as median. So median of a variable can also be called as second quartile Q_2, 5th decile D_5 or 50th percentile P_{50}.

Graphic Method of Location of a Percentile in the Range of a Variable

For graphic location of percentiles refer to Cumulative Frequency Table 2.9 and to ogive, or Figure 3.1

To find the median height or Q_2 in the above example, draw a perpendicular on ogive (Fig. 3.1) from the midpoint of frequency, i.e., 200/2 = 100 and note the point where it

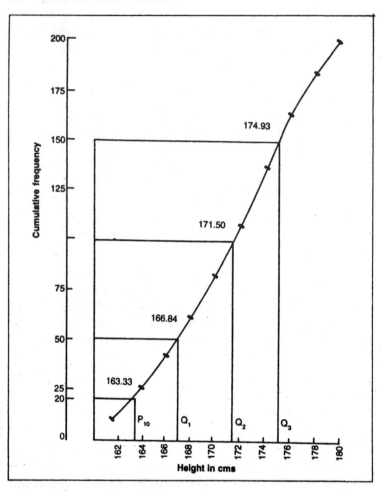

Fig. 3.1: Cumulative frequency diagram showing height values of median (Q_2), first or lower quartile (Q_1), third or upper quartile (Q_3) and tenth percentile (P_{10})

cuts the ogive. Perpendicular from this point on to the baseline locates the median value. It comes to 171.50 cm.

Repeat the same process to find the height of the first or lower quartile Q_1, i.e., of 50th observation (200/4) or the third quartile Q_3, i.e., of 150th observation (200 × 3/4).

Similarly, any percentile value is found, e.g., to find 10th percentile determine the value of 20th observation (200/10).

The values for Q_1, median, Q_3 and 10th percentile are shown in the cumulative frequency diagram (Fig. 3.1).

Conversely from the same diagram, one can find how any given value of a variable divides a distribution into two parts. Thus, what percentage of people are taller or shorter than one's own height can be found. Suppose it is 175 cm. Draw a perpendicular from this height on the baseline on to the ogive (Fig. 3.1) and from ogive to the vertical line. We find 50 persons out of 200 or 25% are taller than 175 cm.

Thus, percentiles are good measures of location.

Arithmetical Method of Finding the Value of any Percentile from the Cumulative Frequency Table

The median, quartile and percentile values of the characteristic can be calculated from the cumulative frequency table. Find the variable group in which the particular observation lies and then raise the lower value of the variable of that group proportionately to the value of that particular observation, with the presumption that the rise in the variable group from the lower value to the higher value is uniform.

For illustration, median, Q_1, Q_3 and 10th percentile are calculated from the Cumulative Frequency Table 2.9.

a. Median (Q_2), i.e., 200/2 or the 100th observation lies in the height group 170–172 cm. The cumulative frequency up to the height less than 170 cm is 81. The frequency rises by 19 from 81 to 100, i.e., median, or middle observation. For 26 observations, the attribute value as per the table, rises by class interval of 2 cm from 170 cm to 172 cm. Therefore, the proportionate rise in the attribute for 19 observations

$$= 2 \times \frac{(100 - 81)}{26} = \frac{2 \times 19}{26} = 1.46$$

Thus, median or second quartile Q_2 value = 170 + 1.46 = 171.46 cm which is almost equal to the graphic value 171.50 cm (Fig. 2.1).

b. First quartile (Q_1), i.e., 200/4 = 50th observation, lies in the group 166–168 cm. Cumulative frequency up to height 166 cm is 42.

$$Q_1 = 166 + 2 \times \frac{(50-42)}{19} \text{ (19 is the group frequency)}$$

$$= 166 + \frac{16}{19} = 166.84 \text{ cm}$$

c. Third quartile (Q_3), i.e., $200 \times 3/4 = 150$th observation, lies in group 174–176 cm. Cumulative frequency up to height 174 is 136.

$$Q_3 = 174 + 2 \times \frac{(150-136)}{30} \text{ (30 is the group frequency)}$$

$$= 174 + \frac{28}{30} = 174.93 \text{ cm}$$

d. Tenth (10th) percentile, i.e., 200/10 or 20th observation, lies in the group 162–164 cm.

$$P_{10} = 162 + 2 \times \frac{(20-10)}{15}$$

$$= 162 + \frac{20}{15}$$

$$= 163.33 \text{ cm.}$$

To find how a particular observation or percentile will divide the entire frequency into two parts of percentages, one has to find in which group of the frequency distribution table this particular observation lies and then raise the frequency value from the lower limit proportionately, again making the presumption that rise is uniform from lower to higher value.

For illustration, let us find how the height 163.33 cm, divides the entire frequency. As seen from the Frequency Distribution Table 2.9, this height lies in cumulative frequency of group 162 cm to 164 cm, i.e., somewhere between 10 and 25, proportional rise of frequency from 10 in this group would be calculated as follows:

Rise in frequency for 2 cm height = 15
Rise in height = 163.33 – 162.00 = 1.33 cm
Rise in frequency for the height 1.33 cm

$$= \frac{15}{2} \times 1.33 = 10$$

Therefore, the number of observations up to the height 163.33 cm = 10 + 10 = 20, which forms the dividing point where 19 observations lie below 20 and 180 above that.

Thus, we make use of the cumulative *frequency distribution table* and not the cumulative frequency polygon.

To summarise, we can locate the size of a given percentile by employing either of the two methods, graphic as well as arithmetic. Conversely we can also find how a given size of an observation divides the frequency distribution into two parts by employing either of the methods.

Application and Uses of Percentiles

1. *Location of a percentile* that divides the frequency distribution into two parts.

Example

The 25th percentile is located in the entire height range as 166.84 cm graphically and arithmetically.

Conversely to find what percentile an observation is.

Example

Height 163.33 cm is found to be 10th percentile graphically and arithmetically. It divides the distribution of 200 heights into 10% and 90%. If a boy's height is less than 163.33 cm, he is short in size for his age, 90% boys are taller than he is

2. *Preparation of a standard percentile* such as quartile Q_1 or median Q_2, etc., for particular age(s), sexes, etc.

Example

Standard median is made use of in the preparation of a weight card, recommended by the Nutrition Subcommittee of Indian Academy of Paediatricians to assess the nutritional status of children in the age group 0–6 years (Fig. 3.2). The same is used in Integrated Child Development Service Scheme (India). Median weights of well-to-do Indian children are found at all months from 1 month to 72 months. By joining the median weights plotted monthwise on the graph paper, 50th percentile for 72 months is drawn and is used as a **standard median**. A comparison between well-to-do

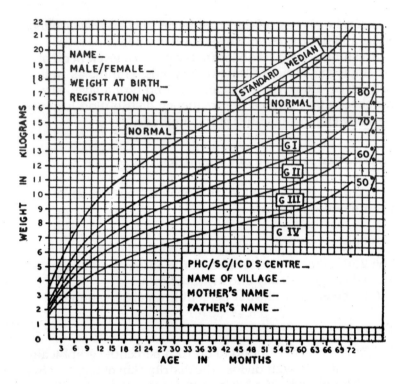

Fig. 3.2: Weight card used for grading weight of children in age group 0–6 years in the ICDS Scheme, India

Indian and American children of corresponding ages revealed that former are as tall and heavy as the latter, boys up to the age of 14 years and girls up to the age of 12 years (V Raghavan K, Darshan Singh and Swaminathan MC, 1971, *Ind J Med Res* 59: 684).

Other lines are drawn 80%, 70%, 60% and 50% of the standard median weights at different ages. In the diagram thus prepared, the vertical axis indicates the weight from 0 to 22 kg and horizontal axis gives the age in months from 1 to 72. Weight at any age is plotted to note between which lines it lies.

Weight for age falling above the first, i.e., 80% line is considered *normal* and that between first line (80% wt) and second line (70% wt) is considered in *malnutrition Grade I*. Weight falling between 70% and 60% lines is in Grade II; between 60% and 50% is in Grade III; and below 50% is considered in Grade IV. Children falling in nutrition Grades III and IV are considered as *severely malnourished*.

3. *Comparison of one percentile value of a variable* of one sample with that of another sample, drawn from the same population or from different population, median, quartile or any other percentiles can be compared.

Example

Cumulative frequency polygons or ogives for one-year-old Indian and Harvard children are drawn taking an identical sample in both the cases (Fig. 3.3). It is seen that Harvard 50th percentile corresponds to Indian 60th percentile. Weight up to 9 kg covers 50% of Harvard children but it

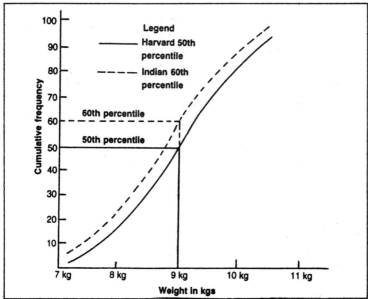

Fig. 3.3: Comparison of weight in Indian and Harvard children at one-year of age by percentiles

covers 60% of the Indian children, showing former children are heavier than the latter.

Thus, size of any percentile of one sample can be compared with other, and deviation found by the difference between the two will be clear at a glance.

4. *To study growth in children*

Examples

1. A cumulative frequency table of heights of male children who have completed 4th, 5th and 6th year of age, is given below:

Age in years	Height in cm up to different percentiles						
	5	10	25	50	75	90	95
4	94.0	96.0	99.8	103.7	106.9	110.5	112.9
5	97.3	99.2	102.6	106.2	110.3	114.2	115.8
6	102.8	104.9	108.1	110.4	115.2	118.4	119.4

Percentiles at different ages indicate growth which is almost uniform when different percentiles are compared. It is about 3 cm from age 4 to 5 and about 4 cm from age 5 to 6. The table also shows how different percentiles divide the distribution of children in two parts.

At age 4, the height 94 cm shows that 5% children were shorter and 95% taller than height 94 cm. Only 5% children are taller than 112.9 cm in this group.

2. The 10th percentile height of 200 children at age 4 was 96 cm. Half of the children (100) were put on vitamins A and D, and the other 100 children were kept as control and given placebo.

After one year, it was found that 10th percentile height was 97.3 cm in control group and 99 cm in experiment group. Thus it was observed that vitamins A and D boosted the height in experiment group (Fig. 3.4).

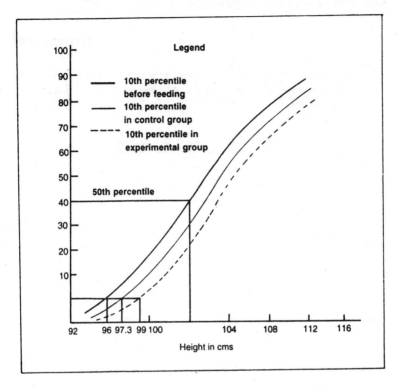

Fig. 3.4: Increase in height after feeding, indicated by 10th percentile in control and experiment groups

5. *As a measure of dispersion* Interquartile and semi-interquartile ranges are sometimes used as measures of dispersion as explained in Chapter 4.

Chapter **4**

Variability and Its Measures

In sciences like physics and chemistry, there is not so much variability as is found in medicine or biology. One chlorine atom is like another but when arranged in biological objects their effects vary. Biological data, quantitative or qualitative, collected by measurement or counting are very variable. No two measurements in man are absolutely equal, not even the means or proportions of two series in health or in disease are equal although we compare like with the like. The cure rate in typhoid with the same drug varies in different patients, mean height of students of the same age group at one place will differ from that at the other place and birth rates of two towns are seldom equal. Apart from the physical characters, the mental qualities such as intelligence quotient, behaviour, and tendency to do wrong or right vary from man to man and group to group.

Variability is essentially a *normal* character. In other words, occurrence of variability is a *biological phenomenon*.

TYPES OF VARIABILITY

There are three main types of variability:
1. Biological variability
2. Real variability
3. Experimental variability.

Biological Variability

Individuals, in similar environments differ when compared as regards, sex, class and other attributes but the difference noted may be small and is said to occur by chance. Such a difference or variability is called **biological variability**. It may be the normal or natural difference in all individuals or groups and may be defined as the one that occurs *within certain accepted biological limits*. The limits of variability of a variable character from the mean and the proportion of a population can be studied by the application of certain *principles of variation* such as normal distribution, range, etc.

The extent of variability, individual or group from a mean or a proportion can be determined by certain measures of deviation or dispersion such as range, interquartile range, standard deviation and standard error. Calculation of such measures usually involves **standard mathematical techniques** which help us to *determine not only the extent of variation* but also whether the difference observed from the central value is due to some factors other than natural. This is the *main* if not the *sole* aim of the study of statistics as a science.

A small variation in birth rates of two cities, say Bombay and Delhi as 26 and 27 per thousand population, respectively, may be considered due to chance; but the birth rate, 27 of Delhi compared to 10 in London would not be taken as a chance variation. Probably it is a true difference due to some human factors interfering with nature such as birth control measures, raising of marriage age or some other social customs, etc.

Difference observed in the mean blood pressures of two groups of individuals beyond chance limits such as between highly placed officers and clerks, may not be inherent or natural but may be unusual, abnormal or even pathological due to external factors such as nature of duty, responsibility, etc.

Reduction observed in typhoid mortality by 1–2% from 28% would be considered due to chloramphenicol and not due to chance or biological reasons.

Classification of Biological Variability

Biological variability may be classified under following heads.

i. *Individual variability* One student's height is 160 cm and that of another of the same age is 170 cm. To find whether their heights are normal or not we find the mean and standard deviation of heights in the population, by taking a large representative sample.

ii. *Periodical variability* The same individuals show variations in temperature, pulse rate, blood pressure, WBC count, blood sugar, urea, cholesterol, etc., at different times of the day, in illness, during rest, after exercise or after meals. The variations can be studied or analysed by applying suitable statistical techniques.

iii. *Class, group or category variability* Height, weight, blood pressure, etc., vary from class to class depending on age, sex, caste, social status or nature of work. Whether the mean height of males is more than females is determined by standard error of difference between two means or by applying unpaired 't' test if the sample is small.

iv. *Sampling variability* To find mean, standard deviation or proportion of certain characters, we do not examine each and every individual in the population but a sample is taken which is much smaller. The values of any sample, therefore, will differ from those of the population and further there will be variability from one sample to another sample. This is a biological variability of samples and is called, *sampling variability* or *sampling error* or *statistical error* which is natural or inevitable. It is measured by finding the *standard error*. Sampling error is a chance variability which can be reduced by increasing the size of the sample but cannot totally be eliminated as explained later under sampling in Chapter 6.

Real Variability

When the difference between two readings, observations or values of classes or samples is more than the defined limits in universe, it is said to be *real*. Then the cause may not be natural or inherent in the man or samples of men but lies in external factors, e.g., statistically significant cure rate may be due to a drug and not by chance. The heights and weights in England are more than those in India perhaps due to better socio-economic conditions. Attack rate in the group, vaccinated against measles is very much less as compared with that in the unvaccinated group because of protection given by vaccination in the former and not due to chance.

Higher rate of coronary disease in bus drivers than that in conductors may not be a biological chance but may be due to strain or tension involved in driving and the sedentary nature of the job of drivers. Difference in the incidence of cancer among smokers and non-smokers may be due to excessive smoking and not due to chance. Higher coronary incidence may be associated with high cholesterol in blood, tension, obesity and such other factors. Standard Error of mean and proportion help to verify and solve such problems. If the variability is found to be real, the offending external factor has to be found and duly dealt with.

Experimental Variability

Error or difference or variation may be due to materials, methods, procedures employed in the study or defects in the techniques involved in the experiment. They are of 3 types—observer, instrumental and sampling.

1. Observer Error

Observer error may be subjective or objective.

i. *Subjective* An interviewer or interrogator may alter some information thereby adding a number of errors while noting human particulars unless trained properly. He may even ask embarrassing questions which the person may not like to answer such as menstrual history, pregnancy, use of family planning methods, sexually transmitted diseases, etc. Some subjects are very keen while others do not wish to give any information. On certain occasions, the situation may not be favourable to seek information.

ii. *Objective* Objective errors may be added by an untrained observer while recording the measurements such as blood pressure, pulse rate, etc. The parallax error, due to different positions of the observer while reading the fluid level in a thermometer or a glass tube or in recording height, weight, etc., is a type of observer error. Disagreement to the extent of 50 per cent has been noted in reading of mass miniature chest skiagrams by radiologists in tuberculosis survey in the recent past. The very act of a beautiful lady doctor or nurse taking pulse may increase the pulse rate of the patient. Biochemical tests may give different results in different hands and so on.

2. *Instrumental Error*

This may be negligible or gross. Defects in weighing machine, height measures, sphygmomanometer, chemical apparatus and other tools may cause undesirable variability or error in observations leading to wrong conclusions and colossal waste of money, time and labour. Hence, all tools employed in a study should be standardised beforehand.

Observer and experimental errors are sometimes called as *non-sampling errors*.

3. *Sampling Defects or Errors of Bias*

A sample drawn should not be biased or too small to draw conclusions. It should be representative and of sufficiently large size to stand statistical tests. Hospital-based studies are mostly biased because the sample of patients under study is drawn from poor, influential or nearby strata of society. Moreover, many come to hospital at a late stage. Birth weight of children born in a private maternity home cannot be compared with those born in a general hospital. One may start with sampling defects if he studies twin birth rates in a hospital when one wants to know the same in general population. Even in field and epidemiological studies, one has to be careful about sampling bias or defects.

Experimental variability due to observer, instrument and sampling defects is not unusual but a common occurrence about which one must be careful in any scientific study so that the bias may be minimised.

MEASURES OF VARIABILITY

Measures of variability of observations help to find how *individual* observations are dispersed around the mean of a large series. They may also be called, *measures of dispersion, variation* or *scatter* as against the averages which are *measures of central tendency*. Measures of variability of individual observations are discussed in this Chapter while measures of variability of samples will be dealt with later, after sampling techniques and probability are explained.

1. *Measures of variability of individual observations*
   i. Range
   ii. Interquartile range
   iii. Mean deviation
   iv. Standard deviation
   v. Coefficient of variation
2. *Measures of variability of samples*
   i. Standard error of mean
   ii. Standard error of difference between two means
   iii. Standard error of proportion
   iv. Standard error of difference between two proportions
   v. Standard error of correlation coefficient
   vi. Standard deviation of regression coefficient.

Range

Variability being a biological characterstic, no single measurement or observation of a variable is considered as an indicator of normality. A range defines the normal limits of a biological characteristic. Some ranges are given below:

Characteristics	Normal limits
Systolic blood pressure	100–140 mm
Diastolic blood pressure	80–90 mm
Fasting blood sugar	80–120 mm
Cholesterol	120–250 mg
Urea	15–40 mg
Uric acid	2–4 mg
Bilirubin	0.2–1.2 mg
Menstrual cycle	21–34 days

Ordinarily observations falling within a particular range are considered normal and those falling outside the normal range are considered as abnormal. Reasons for an observation being lower than the lowest and higher than the highest of the range, have to be searched by the physician. Range for a biological character such as height, haemoglobin, etc., is worked out after measuring the characteristic in large number of healthy persons of the same age, sex, class, etc. It is calculated by finding the difference between the maximum and minimum measurements in the series. Often the two measurements rather than their range are given e.g. in the series of wheal size 3, 1, 6, 10 and 9 mm, the range is 10–1 or 10 to 1 and is given as (1–10) in most of the published works.

Range is the simplest measure of dispersion and is usually employed as a measure of variability in medical practice by one who has little knowledge of statistical methods. It indicates the distance between the lowest and the highest. Though of interest, it is *not a satisfactory measure as it is based only on two extreme values*, ignoring the distribution of all other observations within the extremes. These extreme values vary from study to study, depending upon the size and nature of sample and the type of study. Ordinarily in medical practice, the normal range covers the observations falling in 95% confidence limits.

Just as median and mode are not good measures of central tendency as compared with mean, range too is not a good measure of dispersion as compared with standard deviation. Mean, standard deviation and normal distribution utilise all the available information while median, mode and range do not.

Range does not tell what are the chances of systolic blood pressure being normal when it is 135 mm or 152 mm. This fact is a warning for a medical man not to label a person normal, or pathological on single sign or on single observation whether it is within the range or outside it. He should go by the total picture of a clinical case. There are other measures which describe the variability more accurately.

Semi-interquartile Range (Q)

The range of a variable such as height (Fig. 3.1) between first quartile Q_1 (166.84 cm) and third quartile Q_3 (174.93 cm) is called *interquartile range* (174.93–166.84 = 8.09 cm). Median is the second quartile Q_2 (171.46 cm). Half of this range $\frac{8.09}{2}$ = 4.04 cm is called *semi-interquartile* range or sometimes quartile deviation which is a measure of dispersion around the mean, slightly better than the range (180 – 160 = 20 cm). Q_1 (166.84 cm) divides the frequency into 25% and 75% observations; Q_2 (171.46) divides the frequency into 50% and 50% observations and Q_3 (174.93 cm) divides the frequency into 75% and 25%.

$$Q = \frac{(Q_3 - Q_2) + (Q_2 - Q_1)}{2} = \frac{Q_3 - Q_1}{2}$$

As per heights given above

$$Q = \frac{174.93 - 166.84}{2} = \frac{8.09}{2} = 4.04 \text{ cm}$$

The semi-interquartile range takes into account only the middle half of the data between Q_3 and Q_1. If Q is large, there is greater scatter in the interquartile range. If Q is smaller, there is greater concentration in the middle. If the distribution is symmetrical, $Q_3 - Q_2 = Q_2 - Q_1$, 25% of the observations above and below lie equidistant from the median.

According to the heights,

$$Q_2 - Q_1 = 171.46 - 166.84 = 4.62 \text{ cm}$$

$$Q_3 - Q_2 = 174.93 - 171.46 = 3.47 \text{ cm}$$

Mean distance between Q_1 and Q_3 from the median,

$$= \frac{8.09}{2} = 4.04 \text{ cm}$$

This shows that distribution is not absolutely symmetrical on both sides of the median and 25% observations in the range are not equidistant from the median. It also shows that scatter is more to the left of median (Fig. 3.1).

Mean Deviation

If mean blood pressure of a large representative series is taken, some observations are found above the mean or plus and others are below the mean or minus. On summing up the differences or deviations from the mean, in any distribution, the sum of plus and minus differences will be equal and the net balance will be zero.

To find mean deviation (MD) of observations or measurements from the mean, ignore the sign, add the differences from the mean and divide by the number of observations.

$$MD = \frac{\Sigma |X - \overline{X}|}{n}$$

$|X - \overline{X}|$ within vertical parallel lines or modulus indicates deviations from the mean ignoring negative sign or it is taken as positive. Though simple and easy, mean deviation is not used in statistical analyses being of less mathematical value, particularly in drawing inferences.

Standard Deviation (SD)

Standard deviation is an improvement over mean deviation as a measure of dispersion and is used most commonly in statistical analyses. It is computed by following six steps:
  a. Calculate the mean
  b. Find the difference of each observation from the mean
  c. Square the differences of observations from the mean
  d. Add the squared values to get the sum of squares
  e. Divide this sum by the number of observations minus one to get mean-squared deviation, called **variance** ($\sigma 2$)
  f. Find the square root of this variance to get root-mean squared deviation, called standard deviation. Having squared the original, reverse the step of taking square root.

A large standard deviation shows that the measurements of the frequency distribution are widely spread out from the mean such as 10 mm in case of blood pressure. Small standard deviation means the observations are closely spread in the neighbourhood of mean such as 2 cm in case of height.

Uses of Standard Deviation

1. It summarises the deviations of a large distribution from mean in one figure used as a unit of variation.
2. Indicates whether the variation of difference of an individual from the mean is by chance, i.e., natural or real due to some special reasons.
3. Helps in finding the standard error which determines whether the difference between means of two similar samples is by chance or real.
4. It also helps in finding the suitable size of sample for valid conclusions as explained under sampling in Chapter 6.

Calculation of Standard Deviation

1. First find the mean of series. It is calculated as the sum of observations divided by their number

$$\overline{X} = \frac{\Sigma X}{n}$$

2. Then find the deviations or differences of the individual measurements from the mean $x = X - \overline{X}$.

3. Next find the sum of the squares of deviations or differences of individual measurements from their mean. This can be expressed by the formula,

$$\Sigma(X - \overline{X})^2 = \Sigma \overline{X}^2$$

Where $x = X - \overline{X}$ ...(a)

This can also be calculated by direct method without determining the actual mean by taking differences from an assumed mean 0, or some other convenient working mean (w) by the modified formulae b_1 and b_2 given below, the result will remain the same.

$$\text{Sum of squares} = \Sigma X^2 - \frac{(\Sigma X)^2}{n} \qquad \ldots (b_1)$$

X denotes the size of original observations and the *assumed mean* is 0 (zero).

or

$$\text{Sum of squares} = \Sigma x^2 - \frac{(\Sigma x)^2}{n} \qquad \ldots (b_2)$$

Here, x denotes the differences between the observations and the assumed mean which is a convenient working mean (w).

Deduction of formula (b) from formula (a) is given below.

For illustration the sum of squares is found by using all the 3 formulae for 5 observations of Schick test. The size of red flush in 5 persons was found to be 2, 5, 3, 4 and 1 cm.

Mean or $\overline{X} = \dfrac{\Sigma X}{n} = \dfrac{2+5+3+4+1}{5} = 3$

a.

Size of flush in cm	$X - \overline{X}$ $\overline{X} = 3$ $\overline{X} - 3$	$(X - \overline{X})^2$ or x^2
2	−1	1
5	+2	4
3	0	0
4	+1	1
1	−2	4

Sum of squares of $\Sigma(X - \overline{X})^2 = \Sigma x^2 = 10$

b₁.

Size of flush in cm X	Deviation from assumed mean 0 $(X - 0)$	Squared deviation $(X - 0)^2$
2	2	4
5	5	25
3	3	9
4	4	16
1	1	1
$\Sigma X = 15$		$\Sigma X^2 = 55$

Deduction of formula (b) from (a)

Sum of squares
$= \Sigma(X - \overline{X})^2$
$= \Sigma(X^2 - 2X\overline{X} + \overline{X}^2)$
$= \Sigma X^2 - 2\overline{X}(\Sigma X) + \Sigma \overline{X}^2$
$= \Sigma X^2 - 2\overline{X}(N\overline{X}) + N\overline{X}^2$
$= \Sigma X^2 - 2N\overline{X}^2 + N\overline{X}^2$
$= \Sigma X^2 - N\overline{X}^2$
$= \Sigma X^2 - N\left(\dfrac{\Sigma X}{N}\right)^2$
$= \Sigma X^2 - \dfrac{(\Sigma X)^2}{N}$

Explanations

(a) $\overline{X} = \dfrac{\Sigma X}{N}$ or $N\overline{X} = \Sigma X$

(b) $\Sigma \overline{X}^2 = N\overline{X}^2$

Because sum of squares of means multiplied by number should be same as the number multiplied by mean square. Therefore, $N\overline{X}^2$ is the same as sum of $\overline{X}^2$ takes N times or $\Sigma \overline{X}^2$.

Sum of squares or $\Sigma X^2 = \dfrac{(\Sigma X)^2}{n}$

$= 55 - \dfrac{(15)^2}{5} = 55 - 45 = 10$

This method is conveniently employed when *observations are small in size*

b2.

Size of flush in cm X	Deviation from assumed mean, w X − w = x	Squared deviation x^2
2 (w)	0	0
5	+3	9
3	+1	1
4	+2	4
1	−1	1
Total	$\Sigma x = 5$	$\Sigma x^2 = 15$

Sum of squares $= \Sigma x^2 - \dfrac{(\Sigma x)^2}{n} = 15 - \dfrac{(5)^2}{5} = 10$

This method is applied generally when the observations are large in size.

4. *Now find the variance* (var) which is mean squared deviation, i.e., sum of squares, divided by the number of independent observations. This number is not total but one less than the total number of measurements or observations, (n) in the series, therefore, divide by n − 1. It is also called the *degrees of freedom* in statistical terms and gives an unbiased estimate of variance.

Hence, variance

$= \dfrac{\Sigma(X - \overline{X})^2}{n-1}$ or $\dfrac{\Sigma X^2 - \dfrac{(\Sigma X)^2}{n}}{n-1}$ or $\dfrac{\Sigma x^2 - \dfrac{(\Sigma x)^2}{n}}{n-1}$

For a sample of large size this correction is not applied,

Variance $= \dfrac{\Sigma(X - \overline{X})^2}{n}$

Very often variance is written as Var., or SD^2 or s^2 for sample or σ^2 for universe or population and is made use of in many statistical methods.

5. Lastly, *determine square root of the variance*. That gives the standard deviation which in fact is the square root of the mean squared deviation. It will be in the same units as the original measurements. We had squared the deviations so the square root has to be found.

$$s \text{ or } SD = \sqrt{Var.} = \sqrt{\frac{\Sigma(X - \overline{X})^2}{n - 1}} \text{ by formula (a) or}$$

$$= \sqrt{\frac{\Sigma X^2 - \frac{(\Sigma X)^2}{n}}{n - 1}} \text{ by formula } (b_1) \text{ or}$$

$$= \sqrt{\frac{\Sigma x^2 - \frac{(\Sigma x)^2}{n}}{n - 1}} \text{ by formula } (b_2)$$

Calculation of Standard Deviation(s) in Ungrouped Series

Calculation by formula (a) It is an indirect method because mean has to be calculated first. It is convenient to use when mean is an integral number.

Example

Find the mean respiratory rate per minute and its SD when in 9 cases the rate was found to be 23, 22, 20, 24, 16, 17, 18, 19 and 21.

	Observation (Resp/min) X	Deviation from 20 i.e. mean ($\overline{X}$) $X - \overline{X} = x$	Square of deviation x^2
	23	+3	9
	22	+2	4
	20	0	0
	24	+4	16
	16	−4	16
	17	−3	9
	18	−2	4
	19	−1	1
	21	+1	1
Total	180	0	60

a. $\bar{X} = \dfrac{180}{9} = 20$

b. Variance

$$= \dfrac{\text{Sum of squares of the deviations from the mean}}{n-1}$$

$$s^2 \text{ or Var.} = \dfrac{\Sigma(X-\bar{X})^2}{n-1} = \dfrac{60}{8} = 7.5$$

NB—60 is the sum of the squares of deviations from the mean.

c. s or SD $= \sqrt{s^2}$ or $\sqrt{\text{Var.}} = \sqrt{7.5} = 2.74$

Calculations by formula (b_1) It is a direct method because mean is not to be calculated but is assumed to be 0. It is convenient when observations are small in size. They can directly be squared.

Example

Find SD of the erythrocyte sedimentation rate (ESR), found to be 3, 4, 5, 4, 2, 4, 5 and 3, in 8 normal individuals.

Sum of observations or
$$\Sigma X = 3 + 4 + 5 + 4 + 2 + 4 + 5 + 3 = 30$$

Sum of squares of observations or
$$\Sigma X^2 = 9 + 16 + 25 + 16 + 4 + 16 + 25 + 9 = 120$$

Var. or s^2

$$= \dfrac{\Sigma X^2 - \dfrac{(\Sigma X)^2}{n}}{n-1} = \dfrac{120 - \dfrac{(30)^2}{8}}{8-1} = \dfrac{120 - 112.5}{7} = \dfrac{7.5}{7}$$

$$s = \dfrac{7.5}{7} = \sqrt{1.07} = 1.03$$

Calculation by formula (b_2) It is again a direct method in which mean need not be calculated to find SD but is assumed to be a suitable number. It is convenient when observations are large in size and mean is likely to be expressed to one or two decimal places as happens in most of the cases.

Example

Find SD of incubation period of smallpox in 9 patients where it was found to be 14, 13, 11, 15, 10, 7, 9, 12, and 10.

Observation (IP of smallpox) X	Deviation from assumed mean (w) X–11 = x	Square of deviation x^2
14	+3	9
13	+2	4
11 (w)	0	0
15	+4	16
10	–1	1
7	–4	16
9	–2	4
12	+1	1
10	–1	1
Total 101	2	52

NB—52 is the sum of squared deviations from the assumed mean 11 and not the sum of squares of deviations from the real mean $\left(\dfrac{101}{9} = 11.2\right)$.

The mean, even if found, need not be utilised in calculating SD because fractions will make the calculations laborious.

Sum of squares

Sum of squares of deviations from the assumed mean = $-\dfrac{\text{(Total of deviations from the assumed mean)}^2}{n}$

$= \Sigma x^2 - \dfrac{(\Sigma x)^2}{n}$ (x is deviation and not the observation X)

$= 52 - \dfrac{2^2}{9} = 52 - \dfrac{4}{9} = 52 - 0.44 = 51.56$

$s^2 \text{ or Var.} = \dfrac{\Sigma x^2 - \dfrac{(\Sigma x)^2}{n}}{n-1} = \dfrac{51.56}{8} = 6.445$

$$s \text{ or } SD = \sqrt{\frac{\Sigma x^2 - \frac{(\Sigma x)^2}{n}}{n-1}} = \sqrt{6.445} = 2.54$$

Calculation of Standard Deviation in Grouped Series

The calculation of SD of intelligence quotient (IQ) of 50 boys is given below as an example.

IQ	Mid value of the group	Frequency	Working units	Deviation of group in working units	Squared deviation of group
X		f	χ	f x	f x^2
0-20	10	3	−4	−12	3 × (−4)² = 48
−40	30	4	−3	−12	4 × (−3)² = 36
−60	50	3	−2	−6	3 × (−2)² = 12
−80	70	4	−1	−4	4 × (−1)² = 4
−100	90 w	13	0	0 / −34	0
−120	110	12	1	12	12 × 1² = 12
−140	130	8	2	16	8 × 2² = 32
−160	150	3	3	9 / +37	3 × 3² = 27
Total		50		Σfx = 37 − 34 = 3	Σfx^2 = 171

NB—In working units, difference of 20 is expressed as 1 from the middle value 90 where 0 is placed. x denotes the size of observation, in working units and it is

$$= \frac{X - w}{\text{Class interval}} \text{ or } \frac{X - 90}{20} \text{ in this example}$$

Now mean in working units = $\frac{3}{50}$ = 0.06.

True mean for boys IQ = 90 + 20 × 0.06 = 90 + 1.20 = 91.20. This 90 is the mean value of observations between 81 and 100 opposite which 0 is placed and 20 is the class interval or the unit difference from the assymed mean 90.

$$s^2 \text{ or Var.} = \frac{\Sigma f x^2 - \frac{(\Sigma f x)^2}{n}}{n-1} = \frac{171 - \frac{(3)^2}{50}}{50 - 1} = \frac{171 - 0.18}{49}$$

$$s \text{ or SD} = \sqrt{\text{Var.}} = \sqrt{\frac{171 - 0.18}{49}} = \sqrt{3.48} = 1.86$$

in working units.

True standard deviation of IQ in real units will be $1.86 \times 20 = 37.2$ units. It is multiplied by 20 which is the class interval as done for finding mean in real units.

Thus without finding the mean, SD can be calculated in grouped series also by almost the same steps as for mean, summarised below:

i. Make the frequency table.
ii. Place 0 opposite the middle group (working mean).
iii. Reduce the values to working units as in the case of mean dividing the differences from working mean by class interval.
iv. Just as frequency is multiplied by working units for finding mean, similarly multiply frequency by squares of working units for finding the standard deviation.
v. Then apply the formula and find the mean and SD in working units as before.
vi. Convert this value of SD in real units, multiply SD in working units by the size of class interval.

Coefficient of Variation

It is a measure used to compare relative variability. The variation of the same character in two or more different series has to be compared quite often. It may be of interest to know whether the weight varies more in spleens or in hearts, growth varies more in girls or in boys, variation of pulse rate is more in young or in old, or in students appearing for the examinations and in others not doing so. This is illustrated in example 1.

At other times, the variation of two different characters in one and the same series has to be compared just as height and weight, blood pressure and height, or pulse rate and blood pressure. This is illustrated in example 2.

It compares the variability irrespetive of the units of measurement used in two or more distributions such as height in cm in one and in inches in the other or pulse rate in beats and blood pressure in mm of Hg.

Coefficient of variation (CV) is used to compare the variability of one character in two different groups having different magnitude of values or two charcters in the same group by expressing in percentage.

The coefficient of variation is calculated from standard deviation and mean of the characteristic. The ratio of SD and mean is found in percentages. *Thus SD expressed as percentage of mean is the coefficient of variation.*

Coefficienty of variation, $CV = \frac{SD}{mean} \times 100$, or

$CV = \frac{s}{\bar{X}} \times 100$ where CV stands for coefficient of variation for the series, $\bar{X}$ for the mean of the series and s for the standard deviation.

Examples

1. In two series of adults aged 21 years and children 3 months old following values were obtained for the height. Find which series shows greater variation?

Persons	Mean ht.	SD
Adults	160 cm	10 cm
Children	60 cm	5 cm

CV of adults = $\frac{10}{160} \times 100 = 6.25\%$

CV of children = $\frac{5}{60} \times 100 = 8.33\%$

Thus, we find that heights in children show greater variation than in adults being in the ratio of $\frac{8.33}{6.25}$ or 1.3 : 1.0

2. In a series of boys, the mean systolic blood pressure was 120 and SD was 10. In the same series mean height and SD were 160 cm and 5 cm, respectively. Find which character shows greater variation?

CV of BP = $\frac{10}{120} \times 100 = 8.3\%$

CV of height = $\frac{5}{160} \times 100 = 3.1\%$

Thus, BP is found to be a more variable character than height, 8.3/3.1 = 2.7 times.

Similarly, the weight has been found to be more variable than height.

Chapter 5

Normal Distribution and Normal Curve

So far measures of central tendency, i.e., averages and percentiles and those of variability, viz. range, standard deviation and coefficient of variation have been studied. It has been stated in Chapter 4 that range is not a satisfactory measure of variation whereas standard deviation (SD) summarises the variation of a large distribution, measures the position of an observation with regard to mean, defines the limits of normality and indicates the chances of occurrence of an observation in a population. Though range, averages and percentiles all measure variability or dispersion around the mean but to know how SD does all this, one has to understand what is a *standard distribution* or *normal distribution* of observations in a universe, which also measures variability or observations around the mean.

Out of all the standards or theoretical distributions, normal distribution is most important. Following observations were made on fairly large number of normal persons selected at random.

Observations	Range	Mean	SD	CV
Age at menarche	9–18	14.16	1.13	7.98%
Menstrual cycle in days	21–35	28	2.00	7.14%
Pulse rate in beats/min	55–90	72	3.50	4.86%
Weight at birth in kg	1.8–4.5	3.05	0.39	12.79%
Systolic BP in mmHg	90–150	115	12.00	10.43%

So much of physiological variation, as stated already under range, warns us not to pronounce any isolated observation as abnormal, in otherwise normal person. Still we have to define *normal* or *critical limits* or determine the probable chances of an observation being normal because large deviations from the mean may have unfavourable prognostic value.

If mean menstrual cycle as given above is 28 days and the range is 21 to 35 days, one would like to know how many women will have their cycle between 28 and 30 days or between 26 and 28 days. Similarly, the incubation period of gonorrhoea studied in large number of persons varies from 2 to 8 days with an average of 5. What proportion of exposed persons develop the disease after 5 days or 6 days or before 4 days or between 3 and 4 days, cannot be answered by range and mean.

To determine the chances or probability of an observation or observations being normal and also to determine the proportion of observations that lie between mean and a higher or lower observation we think in terms of frequency distribution as a whole taking into account the mean or central position and dispersion or variability of observations up to the extreme on either side of mean in terms of SD.

When a large number of observations of any variable characteristic such as height, blood pressure and pulse rate are taken at random to make it a representative sample, a frequency distribution table is prepared by keeping group interval small, then it will be seen that:

a. Some observations are above the mean and others are below the mean.
b. If they are arranged in order, deviating towards the extremes from the mean, on plus or minus side, maximum number of frequencies will be seen in the middle around the mean and fewer at the extremes, decreasing smoothly on both sides.
c. Normally, almost half the observations lie above and half below the mean and all observations are symmetrically distributed on each side of the mean.

A distribution of this nature or shape is called *normal distribution* or Gaussian distribution. Normal distribution of observations is a virtue of a large random sample. It is one of the standard distributions in nature. It can be

arithmetically expressed as follows in terms of mean and SD, if they are known:
a. Mean ± 1 SD limits, include 68.27% or roughly 2/3rd of all the observations. Out of the remaining 1/3rd observations, half, i.e., 1/6th will lie below the lower limit (mean – 1 SD) and the other half, i.e., 1/6th will lie above the upper limit (mean + 1 SD). In other words, 32% will lie outside the range, mean ± 1 SD.
b. Mean ± 2 SD limits, include 95.45% of observations while 4.55% of observations will be outside these limits. Similarly, mean ± 1.96 SD limits, include 95% of all observations.
c. Mean ± 3 SD limits include 99.73%.
Mean ± 2.58 SD limits, include 99%.

In other words in any normal distribution it is found that:
a. Observations larger or smaller than mean ± 1 SD are fairly common, forming nearly one-third.
b. Values that differ from the mean by more than twice the standard deviation are rare, being only 4.55%. Their chance of being normal is only 4.55%.
c. Values higher or lower than mean ± 3 SD are very rare, being only 0.27%. Their chance of being normal is 0.27 in 100. Such high values are abnormal or unusual and may be even pathological.

Thus, 6 standard deviations, 3 on either side of the mean cover almost the entire range of a variable character and SD divides the range into 6 equal subranges.

DEMONSTRATION OF A NORMAL DISTRIBUTION

Large quantitative data, collected about any variable has to be summarised by making a frequency distribution table as explained in Chapter 2. This frequency distribution describes how the data are distributed around the mean. Further analysis of data is determined primarily by the shape of this distribution. Table 5.1 gives the frequency distribution of heights of male medical students in a medical college of Gujarat. Mean height and standard deviation were

80 Methods in Biostatistics

found to be 160 cm and 5 cm, respectively. Range is 142.5 to 177.5 cm and class or group interval is 2.5 cm.

Table 5.1: Normal distribution of height

Height in cm	Frequency of each group	Frequency within height limits of				
142.5	3					
145.0	8					
147.5	15					
150.0	45					
152.5	90					
155.0	155	Mean		Mean		Mean
157.5	194	± 1 SD		± 2 SD		± 3 SD
160.0M	195	680		950		995
162.5	136	68%		95%		99%
165.0	93					
167.5	42					
170.0	16					
172.5	6					
175.0–177.5	2					
Mean = 160 cm SD = 5 cm						

1. Range, Mean ± 1 SD = 160 ± 5 or 155 to 165 contains 68% of the observations.
2. Also Mean ± 2 SD = 160 ± 2 × 5 = 160 ± 10 or 150 to 170 will contain 95% of the observations.
3. So also Mean ± 3 SD = 160 ± 3 × 5 or 160 ± 15 or 145 to 175 will contain 99.5% of the observations.
4. The 3 observations lower than mean – 3 SD with heights between 145 cm and 142.5 cm and 2 observations higher than mean + 3 SD with height between 175 cm 177.5 cm fall in the 0.5% group. So their frequency of occurring as normal is only 0.5%, 99.5% chances are that they do not belong to the population under study. Similarly, chances of any observation occurring beyond mean ± 1.96 SD to be normal are 5%, i.e., in 95% cases such observations are abnormal or unusual.

NORMAL CURVE

Histogram of the same frequency distribution of heights, with large number of observations and small class interval, gives a frequency curve which is symmetrical in nature. This is called *normal curve*. The frequency distribution is symmetrical around a single peak, so that mean, median and mode will coincide as in Figure 5.1 below. It builds up gradually from the smallest frequencies at the extremes of classification to the highest frequency at the peak in the middle. This curve is a check on SD to find whether it has been calculated from a random and large sample, representing the universe or not.

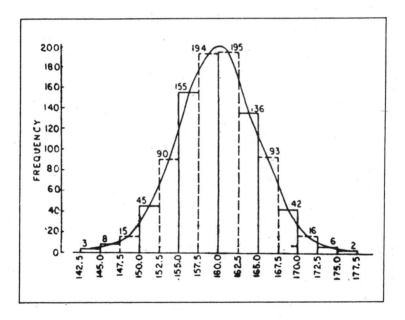

Fig. 5.1: Histogram of 1000 heights in centimetres with normal curve superimposed

The characteristics of a normal curve are:
1. It is bell-shaped or like breast.
2. It is symmetrical.
3. Mean, mode and median coincide.

4. It has two inflections. The central part is convex while at the points of inflection, the curve changes from convexity to concavity. A perpendicular from the point of inflection will cut the base at a distance of one SD from the mean on either side.

Thus, normal distribution and normal curve describe the distribution of frequencies of any variable such as height, weight, blood pressure, bleeding time, clotting time, incubation period, menstrual cycle, etc. Most of these variables follow a normal distribution, expressed arithmetically in terms of mean and standard deviation and shown graphically as a histogram which forms a continuous frequency graph or *frequency curve*, provided the sample is large and group interval is very small. The SD in a normal distribution is the unit of deviation or distance from the mean on the baseline. A perpendicular drawn from the distal end of one SD on to the curve on either side will cut the curve at the points of inflection.

The shape of normal distribution or curve is very useful in practice and makes statistical analysis easy. It tells the probability of occurrence by chance or how often an observation, measured in terms of mean and standard deviation can occur normally in a population (Chapter 7). By normal is meant usual. An unusual observation may have an unfavourable prognosis and be pathological too. Normal and abnormal terms are not used in medical sense.

RELATIVE OR STANDARD NORMAL DEVIATE OR VARIATE (Z)

Deviation from the mean in a normal distribution or curve is called relative or standard normal deviate and is given the symbol Z. It is measured in terms of SDs and indicates how much an observation is bigger or smaller than mean in units of SD. So Z will be a ratio, calculated as below:

$$Z = \frac{\text{Observation} - \text{Mean}}{\text{SD}} = \frac{X - \overline{X}}{\text{SD}}$$

If Z for an observation height 165 cm is plus one, i.e., it is one SD (5 cm) higher than the mean (160 cm), it will cover

50% observations to the left of mean, $\overline{X}$ + 34% to the right of mean (as per normal distribution of observations), i.e., 84%. Only 16% of the total observations will lie above the value mean + 1 SD. In other words, probability of having height above 165 cm is 16% or 0.16 out of one.

Similarly, if Z is 1.5 SD (height 152.5 cm) only 8% (100 – 50 – 34 – 8 = 8) people will have their heights below 150.5 cm.

In the standard normal curve the mean is taken as zero and SD as unity or one. Areas of different values of Z under the curve are tabulated up to two decimal points in Appendix I.

To find the proportion of the individuals who will exceed any particular observation (X) in a standard universe, refer the (Z) value in the table value of the Unit Normal Distribution (Appendix I).

For example, if the ratio comes to 1.5, on referring to the table the proportion comes to 0.0668 out of one. It means 6.68% of individuals will exceed the observation X and 93.32% will not exceed the observation X.

Normal distribution also forms the basis for the tests of significance to be discussed in Chapter 7.

Examples

1. Average weight of baby at birth is 3.05 kg with the SD of 0.39 kg. If the birth weights are normally distributed would you regard:
   a. Wt of 4 kg as abnormal?
   b. Wt of 2.5 kg as normal?
   a. Normal limits of weight at 1.96 SD (3.05 ± 1.96 × 0.39) will be 2.29 kg and 3.81 kg. The weight of 4 kg will be abnormal in more than 95% of cases.
   b. The weight of 2.5 kg lies within the normal limits of 2.29 and 3.81 so it is not taken as abnormal.
2. Menstrual cycle in women following normal distribution has a mean of 28 days and SD of 2 days. How frequently would you expect a menstrual cycle (MC) of·
   a. More than 30 days?
   b. Less than 22 days?

a. As per normal distribution, 68% of the women will have an MC of mean ± 1 SD, i.e., 28 ± 2 or 26 to 30 days, 32% will have longer or shorter MC. Because of symmetry in normal distribution half the women, i.e., 16% women will have longer cycles, i.e., more than 30 days. Calculation may also be done as per Unit Normal Distribution:

$$Z = \frac{X - \bar{x}}{SD} = \frac{30 - 28}{2} = 1$$

Referring to the table of Unit Normal Distribution if Z = 1, the number of women in whom MC will exceed 30 days = 0.1587 out of one, i.e., 15.87%.

b. As per Unit Normal Distribution

$$Z = \frac{22 - 28}{2} = \frac{-6}{2} = -3$$

Corresponding value as per the table is 0.0013. Thus, 0.13% of women will have MC of less than 22 days.

3. Mean height of 500 students is 160 cm and the SD is 5 cm.
   a. What are the chances of heights above 175 cm being normal if height follows normal distribution?
   b. What percentage of boys will have height above 168 cm?
   c. How many of the boys will have height between 168 and 175 cm?

   a. $Z = \dfrac{175 - 160}{5} = \dfrac{15}{5} = 3$

   This corresponds to 0.0013 as per table of Unit Normal Distribution. Thus, only 0.13% of students will have height above 175 cm. The chances of being taller than 175 cm will be 0.13% only.

   b. $Z = \dfrac{168 - 160}{5} = \dfrac{8}{5} = 1.6$

   This corresponds to 0.0548 as per table of Unit Normal Distribution. Thus, number of boys above height of 168 cm will be 5.48% only.

c. The number of boys having height above 168 cm and below 175 cm will be 0.0548 − 0.0013 = 0.0535 out of 1. Thus, only 5.35% of students will have height within the range of 168–175 cm.
4. The pulse rate of healthy males follows a normal distribution with a mean of 72/min and a SD of 3.5/min.
   a. In what percentage of individuals, pulse rate will differ by 2 beats from the mean?
   b. Mark out symmetrically around the mean, the range in which 50% of the individuals will lie?

   a. $Z = \dfrac{X - \bar{x}}{SD} = \dfrac{74 - 72}{3.5} = \dfrac{2}{3.5} = 0.571$

   Referring Z value 0.571 to the Unit Normal Distribution table, the number of individuals in whom pulse rate will exceed 74/min, will be 0.2843 out of 1 or 28.43%. Similarly, number that will have pulse rate less than 70/min will be 28.43%. Thus, in 28.43 + 28.43 = 56.86% of individuals the pulse rate will differ by 2 beats from the mean.

   b. In the normal curve 34% individuals lie on either side of mean up to 1 SD, i.e., 35 beats. So 25% will lie up to $\dfrac{3.5}{34} \times 25 = 2.57$

   Make the area on both sides of mean up to 2.57 SD. Area thus marked will cover 25% + 25% = 50% of the individuals.

ASYMMETRICAL DISTRIBUTIONS

Distributions of different shapes are found in nature. We have already described normal distribution. It forms the basis for practical application for mean and standard deviation as measures of central tendency and variability.

Some distributions are asymmetrical or skewed. They may be skewed to the right (Fig. 5.2) or left depending on whether the long tail of the curve is to the right or left of the peak points. Some of them are bimodal having two peaks as in agewise distribution of deaths in childhood and old age (Fig. 5.3). In such cases, it is probable that two groups of dissimilar

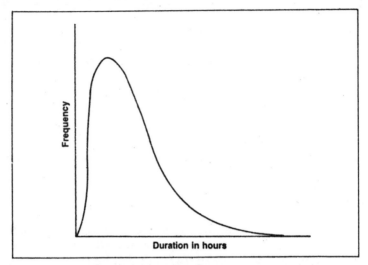

Fig. 5.2: Asymmetrical distribution of duration of labour (1st child) showing skewing to the right

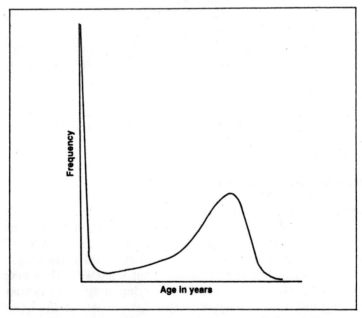

Fig. 5.3: Asymmetrical distribution of deaths according to age showing two modes or peaks in infancy and old age

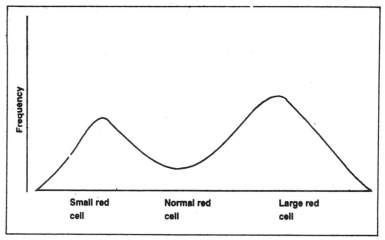

Fig. 5.4: Two types of anaemia giving a bimodal curve

people are mixed together in the population. The sample under study, therefore, is heterogeneous.

Two main types of anaemia—microcytic and macrocytic found in the same population when drawn by size will give a bimodal curve (Fig. 5.4).

Comparison of two or more frequency distributions is easier graphically by converting the table into a histogram or frequency polygon. Centring constants like mean, median and mode of two samples or of a sample and universe will not be in the middle or be the same if the distributions are not normal.

Chapter **6**

Sampling

It is not possible to include each member (sampling unit) of the population in an experimental study or enquiry or examine all the million people of India to find the prevalence of tuberculosis or to test the efficacy of a drug in all the patients suffering from a particular disease. A daily life example is that of cooking rice. A housewife just picks up *a few grains* of rice from the cooking vessel and gets a fairly good idea whether the *entire* lot of rice is fully cooked or it requires more cooking. Further, covering the entire population may be less accurate because a large number of investigators are required to complete the huge task. Their uniformity and correctness may vary, collection will be costly, time consuming and laborious. Because of all such difficulties, we prefer to use an appropriate sampling technique.

In medical studies, the sampling data are collected from a population or universe sufficiently large and representative of the population under study, chosen by a standard sampling technique.

The population or universe must clearly be defined before drawing a sample. For example, population may be an entire group of defined people such as doctors, all members of household, women of 15 to 49 years of age and so on, about whom information is required. It has to be further qualified which doctors, those in service or in general practice, which women, married or unmarried, etc.

A value calculated from a defined population, such as mean (μ), standard deviation (σ), or standard error of mean ($^s\overline{X}$) is called a *parameter*. It is a constant value because it covers all the members of the population. A value calculated from a sample is called a *statistic* such as mean ($\bar{X}$), standard deviation (s) and proportion (p).

The two main objectives of sampling are:
1. *Estimation of population parameters* (mean, proportion, etc.) from the sample statistics.
2. To test the hypothesis about the population from which the sample or samples are drawn.

SAMPLE CHARACTERISTICS

Sample is any part of the population. Large number of samples may be taken from the same population, still all members may not be covered. The composition of samples may vary in size, quality and technique in drawing the sample, thereby their statistics also vary. Inference drawn from a sample refers to the defined population (universe) from which sample or samples are drawn and not to any other population.

Such inferences or conclusions drawn from the sample are applied to the whole population or universe but generalisations are valid, only if the sample is sufficiently large and unbiased, i.e., representative of the entire population from which it is drawn. A representative sample will have its statistics almost equal to the parameters of the entire population. There will still be a chance difference or error of chance, which can be calculated from the representative sample. The differences can be reduced but not eliminated.

There are two main characteristics of a representative sample.
1. Precision which implies the size of the sample
2. Unbiased character

These two qualities help us to fulfil the objectives of samples stated above. Then we can decide, whether the differences observed in values of samples are due to chance or some other factors when compared with population parameters or statistics of another sample.

Precision

Precision depends on the sample size This implies the size, i.e., number in the sample depending on the purpose.

Ordinarily it should not be less than 30. A sample, small in size, is a biased one and should never be depended upon for drawing any conclusions. In typhoid or lobar pneumonia, the case mortality rate or probability of dying after any treatment varied from 20 to 30% before the specific drugs like antibiotics and sulpha were discovered. If a highly qualified doctor, by chance treats only 3 cases, out of those 30% which are going to die, he gets 100% mortality and might be considered inefficient. Another doctor to whom 3 cases coming for treatment from the remaining 70% will be considered more efficient and may even be eulogised. As a matter of fact, neither doctor is to be blamed nor accredited. The wrong conclusion is due to the small size of samples for treatment. If both the doctors were given suitable type and size of samples by applying random technique, mortality in both the cases would have been 25–30%.

While comparing prevalence of diabetes in different occupations in a well conducted diabetes survey, the investigator concluded the prevalence of diabetes amongst soldiers was 66% when only 3 soldiers out of which two had diabetes, entered the study of a mixed population of 5000. He compared diabetes prevalence in soldiers with that in traders who were in large number in the sample. Thus, the size of sample is very vital in any scientific study. Therefore, how large a sample is considered large enough. Normally, the cut off is taken at 30. A sample of size greater than 30 is considered large enough for statistical purposes.

If samples are small in size such as 2, 4, 6, 20, etc., their "means" will be quite different from each other as well as from the population mean, and when the size is 30 or more their "means" will lie closer to each other and will also be in the neighbourhood of population mean. In other words, a small sample lacks precision.

Precision is measured by the formula

$$\text{Precision} = \frac{\sqrt{n}}{s}$$

's', i.e., SD of a sample is the estimate of 'σ', i.e., SD of population. So nearness of s to σ, i.e., precision will be directly proportional to the square root of sample size (n), e.g., if n is increased 4 times, the precision will be double.

Examples

If s is 2 and n is 4, the precision $\frac{\sqrt{n}}{s} = \frac{\sqrt{4}}{2} = \frac{2}{2} = 1$

Assuming that s = 2 and varying the size of n, we calculate the precision as shown below:

If n is 16, precision = $\frac{\sqrt{16}}{2} = \frac{4}{2} = 2$ (precision becomes double if sample size is increased to 4 times)

If n is 36, precision = $\frac{\sqrt{36}}{2} = \frac{6}{2} = 3$ (precision is trebled if n is increased to 9 times)

If n is 64, precision = $\frac{\sqrt{64}}{2} = \frac{8}{2} = 4$ (precision is quadrupled if n is increased to 16 times) and so on.

In case of an experiment, if the results are not decisive about the variation by chance or due to an external factor, increase the size of sample, e.g. if calculated probability (p) is round about 0.05, the conclusion can be confirmed by increasing the size of the sample. To choose the suitable number in order to reduce the sampling error to the minimum and to clear the concept of determining the suitable size of sample to the reader, some of the methods are recommended here.

For Quantitative Data

In such data, we deal with the means of a sample and of the universe. If the SD (σ) in a population is known from the past experience, the size of sample can be determined by the following formulae with the desired *allowable error* (L). At 5% risk the true estimate will lie beyond the allowable error (variation).

Hence, the first step is to decide how large an error due to sampling defects can be tolerated or allowed in the estimates. Such allowable error has to be stated by the investigator.

The second step is to express the allowable error in terms of confidence limits. Suppose L is the allowable error in the sample mean and we are willing to take a 5% chance that the error will exceed L. So we may put:

$$L = \frac{2\sigma}{\sqrt{n}} \text{ or } \sqrt{n} = \frac{2\sigma}{L} \text{ or } n = \frac{4\sigma^2}{L^2}$$

If $L = 1$ and $\sigma^2 = 25$, $n = \frac{4 \times 25}{1^2} = 100$

In such cases, the investigator may start with an assumed SD and the allowable error specified by the experimenter. In case SD is not known, preliminary investigation or a pilot survey may have to be carried out to estimate the population SD.

Examples

1. Mean pulse rate of a population is believed to be 70 per minute with a standard deviation of 8 beats. Calculate the minimum size of the sample to verify this, if allowable error $L = \pm 1$ beat at 5% risk.

$$n = \frac{4\sigma^2}{L^2} = \frac{4 \times 8 \times 8}{1 \times 1} = 256$$

(*Source*: Snedecor GW, fifth edition p. 502)
If $L = \pm 2$ beats with 5% risk

$$n = \frac{4 \times 8 \times 8}{2 \times 2} = 64$$

If L is less, n will be more, i.e., larger the sample size, lesser will be the error.

2. Mean systolic blood pressure in one college students was found to be 120 with SD of 10. Calculate the minimum size of the sample to verify the result if allowable error is ± 2 at 5% risk.

$$n = \frac{4\sigma}{L^2} = \frac{4 \times 10 \times 10}{2 \times 2} = 100$$

For Qualitatitve Data

In such data, we deal with proportions such as morbidity rates and cure rates. For finding the suitable size of the sample, the assumption usually made is that the allowable error does not exceed 10% or 20% of the positive character. The size can be calculated by the following formula with a desired allowable error (L) at 5% risk that the true estimate will not exceed allowable error by 10% or 20% of 'p'

$$n = \frac{4pq}{L^2}$$

where 'p' is the positive character, q = 1 – p and L = allowable error, 10% or 20% of 'p'.

Examples

1. Incidence rate in the last influenza epidemic was found to be 50 per thousand (5%) of the population exposed. What should be the size of sample to find incidence rate in the current epidemic if allowable error is 10% and 20%?

$$n = \frac{4pq}{L^2} \quad p = 5\%, q = 95\%$$

if L = 10% of p = $5 \times \frac{10}{100}$ = 0.5%

$$n = \frac{4pq}{L^2} = \frac{4 \times 5 \times 95}{0.5} \times 0.5 = 7600$$

If L = 20% of p = $5 \times \frac{20}{100}$ = 1%,

$$n = \frac{4pq}{L^2} = \frac{4 \times 5 \times 95}{1 \times 1} = 1900$$

So larger the permissible error, the smaller will be the size of sample required for both types of data.

2. Hookworm prevalence rate was 30% before the specific treatment and adoption of other measures. Calculate the size of the sample required to find the prevalence rate now if allowable error is 10% and 20%.

If L = 10% of p = $\frac{30 \times 10}{100}$ = 3

At 5% risk, $n = \dfrac{4pq}{L^2} = \dfrac{4 \times 30 \times 70}{3 \times 3} = 933.3$ (or 934 to round off the size.

If L = 20% of p, $= \dfrac{30 \times 20}{100} = 6$

At 5% risk $n = \dfrac{4pq}{L^2}$

$= \dfrac{4 \times 30 \times 70}{6 \times 6} = 233.3$ (or 234 to round off the size)

Thus, if we allow a small error, the required sample size will be much larger as compared to one when the allowable error is increased.

Unbiased Character

Sample bias: Bias comes in when the samples from a population are not chosen at random or samples are not drawn from similar populations. Bias may creep in due to non-sampling errors already discussed under experimental variability in Chapter 4. Bias due to sampling defects alone is discussed here. When there is bias, the statistics of samples like mean ($\bar{X}$) will be away from the population parameter (μ).

The mean weight of 50 babies at birth would be nearly the mean weight of the total babies born in the population. If these babies were born in a private maternity home where only women from the affluent families come for delivery, then they form a biased or selected sample of all babies. They may give an unbiased estimate of mean baby weight in well-to-do class but would seriously overestimate the mean weight of babies in a mixed population.

The cure rate of cases of typhoid or diphtheria treated by general practitioners, cannot be compared with that of hospital cases. Patients treated by private practitioners are mostly those who are in early stage of disease, conscientious, educated, co-operating and from higher socio-economic strata as compared with those treated in government hospitals who come in advanced stages of disease and belong to lower socio-economic strata. Both types if chosen for generalisation, are biased or selected samples. Prevalence rate of tuberculosis in a well to-do locality or in slums may not give the correct picture of tuberculosis prevalence in India.

Selection or bias should be avoided in any scientific study when we compare like with the like. The age, sex, social status, stage of disease should be same in both the samples, but sometimes it may be deliberate, e.g. we specially want to know the spleen rate in secondary school children only, or the mean blood pressure of people above 50 facing high responsibility or the pulse rate of infants and so on. In such cases, generalisation is to be made about a particular group to compare with another group which is different such as coronary disease in drivers and conductors. Here, the primary objective is to compare like with unlike.

Selection or bias may creep in unconsciously and go unnoticed, hence, it has to carefully be avoided while conducting an experiment, e.g. in enquiries by questionnaire, selection or bias must be suspected. Too keen and intelligent persons might volunteer information and even exaggerate, while those not keen and some with particular habits such as smoking or with disease as gonorrhoea, might hide the facts or not reply at all. Unconsciously, sometimes we compare the result of treatment in one group of tuberculosis cases with that of the other, disregarding factors like one or two lungs affected, extent of the disease, age, sex, class, occupation, etc.

In some studies, the effort is made to compare like with the like, but because of unavoidable bias or selection bias, due to number of disturbing factors, one sample may not be similar to the other. Thus, the results may be affected and wrong conclusions drawn, e.g. if the development of babies on breast feeding is compared with those on bottle feeding, it is difficult to get similar samples. Many mothers start supplements in various forms and disturb similarity among the groups.

It is often difficult to rule out bias in retrospective studies and in studies where subjective observations (information given by subjects in the sample) are made. It is easy to avoid bias in prospective studies and in experiment where objective observations (those made by investigator) are made. Subjective bias is ruled out by single or double blind trials.

Wherever possible, a control is a must. Number and type of subjects included in the experiment and technique used for selection should be same for control and experimental groups and decided before choosing the subjects in either group.

Following sampling techniques are employed to choose an unbiased sample. If these techniques are employed, the chance error can be calculated and the probability or relative frequency of getting different results from sample to sample or from sample to those of population can be determined. Hence, samples, thus selected are called *probability samples*. Any unit or member of such samples has a definite probability or chance of being included, e.g. probability of drawing spade ace or any specific card in a pack of playing cards is one in 52 in one draw and of any ace, the probability is 13 in one draw. Probability will be explained in detail in Chapter 7.

SAMPLING TECHNIQUES

Simple Random Sampling

The method is applicable when the population is small, homogeneous and readily available such as patients coming to hospital or lying in the wards. It is used in experimental medicine or clinical trials like testing the efficacy of a particular drug. The principle here is that every unit of the population has an equal chance of being selected. Hence, this method is also sometimes called, 'unrestricted random sampling'.

The sample may be drawn unit by unit, either by numbering the units such as persons, families or households of a particular population on the cards or from the published tables of random numbers. To ensure randomness of selection, one may adopt either lottery method or refer to table of random numbers.

Lottery Method

Suppose, 10 patients are to be put on a trial out of the 100 available. Note the serial number of patients on 100 cards and shuffle them well. Draw out one and note the number. Replace the card drawn, reshuffle and draw the second card. Repeat the process till 10 numbers are drawn. Reject the cards that are drawn for second time. The 10 cards drawn thus will indicate the patient's number to be put on trial and the 10 patients selected in this manner form the random sample. Similar procedure may be followed for selecting the control if need be. This is sampling with replacement.

Table of Random Number Method

The other common method of drawing the sample is by making use of the published tables of random numbers. To draw a sample of 10 out of 100 with the help of table of random numbers we may refer to Appendix VI. First give serial numbers to all the 100 patients in the above example at random, starting from any patient. This reduces the bias at the very start. Total number of patients or population from which desired number 10 is to be chosen is 100, a 3 digit number.

Numbers, less than 100 may be chosen as they are, while those higher than 100 may be divided by 100 and the remainders may be note as the numbers chosen for the sample. Number higher than 100 could be rejected too, making use of the rows below. Thus, 10 numbers for the sample are chosen. We can start from any row or column or even diagonally.

Example

In this example, start at the 11th, 12th and 13th columns from the first row and move downwards row-wise to get following 10 numbers:

369	495
428	572
565	169
969	786
385	094

The number selected for the sample will be 094, i.e., 94th patient of the series and (369/100, remainder 69) 69, 28, 65, 69, 85, 95, 72, 69 and 86. The number 69 has appeared thrice, hence, numbers 969 and 169 have to be rejected and two more taken from the subsequent rows, i.e., 441 and 841 on the table in the same columns. Thus, 9th patient will be 41st and 10th patient will be remainder of the next 3 digits. i.e., 29 of 829 in the 13th row (rejecting 841 because 41 appears again). Thus the numbers to be put on trial will be 69, 28, 65, 85, 95, 72, 86, 94, 41 and 29. This way the sample of a desired size can be drawn from any size of the universe or population, e.g. to select a sample of 150 out of 15000 population start with any 5 digits row-wise or columnwise such as 03474 in the first row of Appendix V. Every time a number less than the population has to be chosen such as 03474 and higher number has to be divided by the population, such as 97742 taking the remainder 7742 as the number selected. Continue this process till 150 numbers are chosen.

Systematic Sampling

This method is popularly used in those cases when a complete list of population from which sample is to be drawn, is available. It is more often applied to field studies when the population is large, scattered and not homogeneous. Systematic procedure is followed to choose a sample by taking every Kth house or patient where K refers to the sample interval, which is calculated by the formula:

$$K = \frac{\text{Total population}}{\text{Sample size desired}}$$

e.g. if 10% sample is to be taken out of one thousand patients,

$$K = \frac{1000}{10\% \text{ of } 1000} = 10$$

One random number is found by pulling out one card after shuffling, out of 10 cards serially numbered 1 to 10. Supposing it is 6, then the sample will consist of units with sample numbers 6, 6 + 10 = 16, 16 + 10 = 26, 26 + 10 = 36 and so on. Examine every 10th house after the 6th house.

such as was done to assess incidence of influenza in one epidemic in a large city like Bombay.

If 20% sample is to be taken, $K = \frac{100}{20\% \text{ of } 100} = 5$, examine every 5th case starting with the random number such as 2. subsequent numbers will be 2 + 5, 7 + 5, 12 + 5, i.e., 7, 12, 17, 22 and so on.

If 5% sample is to be taken $K = \frac{100}{5\% \text{ of } 100} = 20$, then every 20th number has to be taken starting with the random number such as 2, the subsequent number will be 22, 42, 62, 82 and so on. Alternate cases of a disease may be taken in the trial of a drug then $K = \frac{100}{0.5\% \text{ of } 100} = 2$

Merits

1. The systematic design is simple, convenient to adopt.
2. The time and labour involved in the collection of sample is relatively small.
3. If the population is sufficiently large, homogeneous and each unit is numbered, this method can yield accurate results.

Stratified Sampling

This method is followed when the population is not homogeneous. The population under study is first divided into homogeneous groups or classes called strata and the sample is drawn from each stratum at random in proportion to its size.

It is a method of sampling for giving representation to all strata of society or population such as selecting sample from defined areas, classes, ages, sexes, etc. This technique gives more representative sample than simple random sampling in a given large population, e.g. the entire population may be divided into five socio-economic groups or strata. Out of large groups 10% may be selected for study and 50% or even more from the smaller strata so that no subsample is less than 30 in size.

Merits

1. Proportionate representative sample from each strata is secured.
2. It gives greater accuracy.

Multistage Sampling

As the name implies, this method refers to the sampling procedures carried out in several stages using random sampling techniques.

This is employed in large country surveys. In the first stage, random numbers of districts are chosen in all the states, followed by random numbers of talukas, villages and units, respectively, e.g. for hookworm survey in a district, choose 10% villages in the talukas and then examine stools of all persons in every 10th house.

Merits

1. It introduces flexibility in sampling, which is lacking in other techniques.
2. It enables the use of existing division and subdivision which saves extra labour.

Cluster Sampling

A cluster is a randomly selected group. This method is used when units of population are natural groups or clusters such as villages, wards, blocks, slums of a town, factories, workshops or children of a school, etc. Fortunately, the technique of cluster sampling allows small number of the target population to be sampled while the data provided is statistically valid at 95% confidence limits (10% variation). From the chosen clusters, 30 in number, the entire population is surveyed. Cluster sampling gives a higher standard error but the data collection in this method is simpler and involves less time and cost than in other sampling techniques.

As per module approved by WHO, it is most often used to evaluate vaccination coverage in Expanded Programme of Immunization (EPI) and Universal Immunization

Programme (UIP), where only 210 children, taking 7 from each cluster in the age group 12-23 months are to be examined.

Identification of Clusters for Collection of Data

1. List all cities, towns, villages and wards of cities with their population falling in the target area under study for evaluation.
2. Calculate cumulative population and divide the same by 30. This gives the SAMPLING INTERVAL.
3. Select a random number less than or equal to sampling interval having same number of digits. This forms the first cluster.
4. Random number plus sampling inverval gives the population of 2nd cluster.
5. Second cluster + sampling interval = 3rd cluster.
   3rd cluster + sampling interval = 4th cluster and so on.
   Last or 30th cluster = 29th cluster + sampling interval.
   All houses with population are numbered.

The first house for survey in each cluster should randomly be selected with the help of random number table or number on a currency note. Before starting house to house survey, define the age group and item you wish to study, e.g. children in age group 12-17 months, fully vaccinated. They have had 3 DPT, 3 Polio, 1 BCG and 1 measles vaccination. Now survey houses, starting with the selected house till you get 7 children fully vaccinated. Thus, 210 such children will be found in 30 clusters.

$$\text{Coverage} = \frac{210}{\text{Eligible children of surveyed houses}} \times 100$$

Multiphase Sampling

In this method, part of the information is collected from the whole sample and part from the subsample. In a tuberculosis survey, physical examination or Mantoux test may be done in all cases of the sample in the first phase; in the second phase X-ray of the chest may be done in Mantoux positive cases and in those with clinical symptoms, while sputum may be examined in X-ray positive cases in the third phase only. Number in the subsamples in 2nd and 3rd phases will become successively smaller and smaller. Survey by such procedure will be less costly, less laborious and more purposeful.

Sometimes two or more independent samples may be taken by different survey teams to compare the results. Such samples may overlap to some extent or be exclusive of each other.

There are other methods needed in different situations for which reader may consult some other books on sampling.

SUMMARY

To summarise and clarify the concept of sampling, the essential charateristics of a representative sample are:
1. Its size should be 30 or more—larger the sample, lesser would be the error due to chance.
2. It should randomly be selected by an appropriate sampling technique so that each member of the population has an equal opportunity of being selected.

The rules for selection should be independent of the observations to be made. If the sample represents the population, its values or statistics like mean ($\bar{X}$), SD (s) or proportion (p) will not differ significantly from population parameters or true value μ, σ or P, respectively. Difference found will not exceed the chance error.

If you pick up one apple of any size from a heap, it does not represent the entire lot. It may be of big, small or medium size. Its size and weight may not be equal to the mean size and weight of the lot. If you pick up 30 at random after mixing all sizes, you will get a representative sample of 30 apples representing the entire heap (sufficiently large and unbiased). The seller can then fix the rate of 30 apples from any side of heap, not allowing the customer to choose.

Chapter 7

Probability (Chance)

Probability may be defined as the *relative frequency or probable chances of occurrence* with which an event is expected to occur on an average such as of giving birth to a boy in the first pregnancy, chances of one drug being better than the other, likelihood of a particular blood pressure on the higher side of the range being normal and so on. In other words, it denotes the *relative frequency* or *odds with which an event is expected to occur* on an average or in the long run such as a *twin pregnancy* will occur once in *80 pregnancies* or a child with blood group *Rh negative* will be born *once in ten births* and so on.

The main purpose of selecting a representative sample (Chapter 6) is to know the *probability (relative frequency)* of occurrence of single or group of observations in a normal distribution of any biological variable. We would also like to know the probability (chance) of occurrence of sample values (means and proportions) by chance so that sample results can be compared with those of population. Chance variability can then be ruled out and inference about the population can be drawn. To do all this, probability of occurrence of biological happenings or events in universe must be well understood.

An element of uncertainty is associated with every conclusion because information on all happenings is not available. It is inevitable beyond forecast. This uncertainty is numerically expressed as *probability*. It measures the relative frequency of a particular event happening by chance in the long run. Inferences or conclusions drawn after various statistical analyses are based on theory of probability.

Probability is usually expressed by the symbol 'p'. It ranges from zero (0) to one (1). When p = 0, it means there is no chance of an event happening or its occurrence is impossible. Chances of survival after rabies are zero or nil. If p = 1, it means the chances of an event happening are 100%, i.e., it is inevitable or a case of certainty such as death for any living being. Probability of survival after sandfly fever is 100%.

If the probability of an event happening in a sample is p and that of not happening is denoted by the symbol q, then
$$q = 1 - p \text{ or } p + q = 1$$
One can estimate the probability by logic alone, e.g. the probability p of drawing any one of the 4 aces in one attempt from a pack of 52 cards is $\frac{4}{52} = \frac{1}{13}$

and of not drawing is $q = 1 - \frac{1}{13} = \frac{12}{13}$

Chances of getting head or tail in one toss or of getting male or female child in one pregnancy are fifty-fifty or half and half, i.e., $p = \frac{1}{2}$ and $q = \frac{1}{2}$ where 'p' denotes the probability of getting a male child and 'q' of not getting a male child or getting a female child. Arithmetically, we calculate the probability (p) or chances of occurrence of a positive event by the formula:

$$p = \frac{\text{Number of events occurring}}{\text{Total number of trials}}$$

If a surgeon transplants kidney in 200 cases and succeeds in 80 cases then probability of survival after operation is calculated as:

$$p = \frac{\text{Number of survivals after the operation}}{\text{Total number of patients operated}}$$

$$= \frac{80}{200} = \frac{2}{5} = 0.4$$

'q' or probability of not surviving or dying

$$= \frac{\text{Number of patients died}}{\text{Number of patients operated}} = \frac{120}{200} = \frac{3}{5} = 0.6$$

Probability of getting 6 in one throw of dice is $\frac{1}{6}$ and of getting other than 6, i.e., 1, 2, 3, 4, or 5 will be $1 - \frac{1}{6} = \frac{5}{6}$. Total events that can happen are six and '6' will be the only one favourable event out of that.

If twins are born once in 80 different pregnancies, then p for birth of twins = $\frac{1}{80}$ and the probability for single birth $q = 1 - \frac{1}{80} = \frac{79}{80}$.

If probability of being Rh –ve is $\frac{1}{10}$, then of being Rh +ve will be $1 - \frac{1}{10} = \frac{9}{10}$.

LAWS OF PROBABILITY

It is very important to have a clear concept of probability as it provides the *basis for all the tests of significance*. It is estimated usually on the basis of following five laws of probability, normal curve and tables.
1. Addition law of probability
2. Multiplication law of probability
3. Binomial law of probability distribution
4. Probability (chances) from shape of normal distribution or normal curve
5. Probability of calculated values from tables.

Addition Law of Probability

If one event excludes the probability of occurrence of the other specified event or events, the events are called *mutually exclusive*. Getting head excludes the possibility of getting tail on tossing a coin, birth of a male excludes birth of a female, throw of 2 excludes other five events, i.e., 1, 3, 4, 5 and 6 in one throw of dice, Rh –ve birth excludes birth of an Rh +ve baby, dying excludes survival and so on.

An event will occur in one of the several ways. The birth will be of a male or a female baby. Blood groups will be A, B, O or AB; WBCs will be polymorph, lymphocytes, monocytes eosinophils or basophils; card drawn once will be an ace or one of the other 12 cards and so on. They all have an individual probability or relative frequency of occurrence. Total probability will be equal to the sum of individual probabilities provided the events are mutually exclusive.

Thus, **mutually exclusive** events follow the addition law of probability.

If the number of mutually exclusive events are n and p_1 in the individual probability then total probability, P, is calculated as

$$P = p_1 + p_2, ..., + p_n = 1$$

The word '**or**' is there when addition law is applied, e.g. getting Rh –ve or Rh +ve child, a drug will cure or relieve or have no effect on a disease, result of a dice thrown once will be 1 or 2 or 3 or 4 or 5 or 6 and so on. Probability of getting only '2' will be $\frac{1}{6}$ and that of getting '6' is $\frac{1}{6}$. The total probability of getting 2 or 6 in one throw will be $\frac{1}{6} + \frac{1}{6} = \frac{2}{6} = \frac{1}{3}$.

Probability of getting a male child is $\frac{1}{2}$ and total of getting male or female = $\frac{1}{2} + \frac{1}{2} = 1$. In one cut, chance of getting jack of hearts is $\frac{1}{52}$ and of getting any of the four jacks will be

$$\frac{1}{52} + \frac{1}{52} + \frac{1}{52} + \frac{1}{52} = \frac{4}{52} = \frac{1}{13}$$

In the same way, probability of getting any jack or king in one draw will be $\frac{1}{13} + \frac{1}{13} = \frac{2}{13}$ while that of getting only jack of spades or king of spades will be $\frac{1}{52} + \frac{1}{52} = \frac{2}{52} = \frac{1}{26}$

Total probability of getting Rh +ve or Rh −ve child will be $\frac{9}{10} + \frac{1}{10} = 1$.

Multiplication Law of Probability

This law is applied to two or more events occurring together but they must not be associated, i.e., must be independent of each other. The word 'and' is used in between the events such as 2 and 5 or 5 and 2 in two throws of dice.

A dice is thrown twice in succession, what will be the probability of getting 5 and 2 or 2 and 5?

In the first case, probability of getting 5 in the first throw is 1/6 and 2 in the second throw is 1/6, so the probability of getting 5 in the first throw and 2 in the second would be $1/6 \times 1/6 = 1/36$.

In the second case, the probability of getting 2 in the first throw and 5 in the second throw would be again $1/6 \times 1/6 = 1/36$.

It confirms that sequence is immaterial.

If sequence of 5 and 2 or 2 and 5 is not taken into account, the probability of getting either in two throws is $1/36 + 1/36 + 2/36 = 1/18$.

Multiplication and addition laws both have to be applied in such cases, both the words 'and' and 'or' are to be used.

Sex of birth and Rh factor are independent events and occur in any child.

What will be the probability of a child being male and Rh +ve?

Probability of child being male, $p_1 = \frac{1}{2}$

Probability of child being Rh +ve, $p_2 = \frac{9}{10}$

Probability of a single birth $p_3 = \frac{79}{80}$

Probability of a child being male and Rh +ve = $p_1 \times p_2 = \frac{1}{2} \times \frac{9}{10} = \frac{9}{20}$

In the same way, probability of a female, Rh +ve and single child being born

$$p_1 \times p_2 \times p_3 = \frac{1}{2} \times \frac{9}{10} \times \frac{79}{80} = \frac{711}{1600}$$

Out of 100 persons, 10 smoke heavily and probability of getting cancer in heavy smokers is one in 100 (hypothetical data). What are the chances of being a heavy smoker and getting lung cancer? 10/100 × 1/100 = 1/1000, it is not correct.

Here, law of multiplication is not applicable because smoking and lung cancer are associated and not mutually exclusive.

Probability of being colour blind is one in 12 and of being male is 1/2. Here, too, one cannot use multiplication law (1/2 × 1/12 = 1/24) as colour blindness and male sex are associated. Colour blindness is usually found in male children only.

Application of law of multiplication when events occur in more than one sequence such as 2 and 2 or 2 and 6 or 6 and 2 or 6 and 6 in two throws of dice may be extended to find the probability of sex of second, third or any subsequent child after the first.

Binomial Law of Probability

When *two children* are born one after the other, the possible sequences will be any of the following four:

1st issue	2nd issue		Sequences	Probability
M	M	(1)	M and M	$\frac{1}{2} \times \frac{1}{2} = \frac{1}{4}$
M	F	(2)	M and F	$\frac{1}{2} \times \frac{1}{2} = \frac{1}{4}$
F	M	(3)	F and M	$\frac{1}{2} \times \frac{1}{2} = \frac{1}{4}$
F	F	(4)	F and F	$\frac{1}{2} \times \frac{1}{2} = \frac{1}{4}$

Chances of getting 2 males = $\frac{1}{4}$ = 25%

Chances of getting 2 females = $\frac{1}{4}$ = 25%

Chances of getting one of either sex will be total of second and third sequence = $\frac{1}{4} + \frac{1}{4} = \frac{1}{2}$ = 50%.

So, if a female child is born first and a second child is desired, the probability of the second issue being male will be 75% and of its being female 25%.

The probability of sequences (2) and (3)
$$= \frac{1}{2} \times \frac{1}{2} + \frac{1}{2} \times \frac{1}{2} = \frac{1}{4} + \frac{1}{4} = \frac{2}{4} = \frac{1}{2} = 50\%.$$

Probability of 2 F = 25%. So that of second child being male = 100 − 25% = 75%.

Similarly, when *three children* are born the possible sequences will be any one of the 8 given below:

Issue				Sequences	Probability
1st	2nd	3rd			
M	M	M		(1) MMM	$\frac{1}{2} \times \frac{1}{2} \times \frac{1}{2} = \frac{1}{8}$
M	M	F		(2) MMF	$\frac{1}{2} \times \frac{1}{2} \times \frac{1}{2} = \frac{1}{8}$
M	F	M		(3) MFM	$\frac{1}{2} \times \frac{1}{2} \times \frac{1}{2} = \frac{1}{8}$
M	F	F		(4) MFF	$\frac{1}{2} \times \frac{1}{2} \times \frac{1}{2} = \frac{1}{8}$
F	M	M		(5) FMM	$\frac{1}{2} \times \frac{1}{2} \times \frac{1}{2} = \frac{1}{8}$
F	M	F		(6) FMF	$\frac{1}{2} \times \frac{1}{2} \times \frac{1}{2} = \frac{1}{8}$
F	F	M		(7) FFM	$\frac{1}{2} \times \frac{1}{2} \times \frac{1}{2} = \frac{1}{8}$
F	F	F		(8) FFF	$\frac{1}{2} \times \frac{1}{2} \times \frac{1}{2} = \frac{1}{8}$

Probability of getting 3 males (seq 1) = $\frac{1}{2} \times \frac{1}{2} \times \frac{1}{2} = \frac{1}{8}$

Probability of getting 3 females (seq 8) = $\frac{1}{2} \times \frac{1}{2} \times \frac{1}{2} = \frac{1}{8}$

Probability of getting 2 males and 1 female = $\frac{1}{8} + \frac{1}{3} + \frac{1}{8}$

(sum of seq nos 2, 3 and 5) = $\frac{3}{8}$

Probability of getting 1 male and 2 females = $\frac{1}{8} + \frac{1}{8} + \frac{1}{8}$

(sum of seq nos 4, 6 and 7) = $\frac{3}{8}$.

Total probability = $\frac{1}{8} + \frac{1}{8} + \frac{3}{8} + \frac{3}{8} = \frac{8}{8} = 1$

In case of 3 siblings, the proportion of percentages of four will be:

Combinations	Probability		Percentage
1. Three males	0.125	or	12.5
2. Three females	0.125	or	12.5
3. Two males one female	0.375	or	37.5
4. Two females one male	0.375	or	37.5
Total	1.000		100

The chance of getting 2 of one sex and one of the opposite sex = 37.5 + 37.5 = 75%.

But if the first 2 are female children and the third is desired to be male, the chances are 75 + 12.5 = 87.5%, because probability of all three being females is 12.5% only. i.e., 100 − 12.5 = 87.5%.

The foregoing discussion on probability is governed by the Binomial Probability Distribution.

Binomial law of probability distribution This is formed by the terms of the expansion of the binomial expression $(p + q)^n$, where n = sample size or number of events such as births, throws, tosses or persons selected at random for whom the probability is to be worked out, p = the probability of a 'success', q = probability of a 'failure', and $p + q = 1$.

Examples

1. When n = 2, the terms of the expansion of $(p + q)^2$ are p^2, $2pq$ and q^2.
2. When n = 4, the terms of the expansion of $(p + q)^4$ are p^4, $4p^3q$, $6p^2q^2$, $4pq^3$ and q^4.

Values of p and q are found from the population percentage, i.e., chance of getting a boy (value of p) or a girl (value of q) when only one child is born, is found by observing large number of births in a universe. It may be 51% for boy (p) and 49% for girl (q). So p = 0.51 and q = (1 − 0.51) = 0.49. Substitute these values and find the probability.

Examples

1. What are the chances of getting any combination, i.e., 2 boys, 2 girls or one boy and one girl when number of pregnancies is 2?

   The probability of these 3 outcomes can be calculated by the formula $(p + q)^2 = p^2 + q^2 + 2pq$, where p^2 would mean probability of getting 2 boys, q^2 of 2 girls and $2pq$ of one boy and one girl.

   Substitute the probability p and q in the above formula, i.e., 0.51 and 0.49.

   $(p + q)^2 = (0.51)^2 + (0.49)^2 + (2 \times 0.51 \times 0.49)$
   $= 0.2601 + 0.2401 + 0.4998$.

   Probability of proportional chances of getting 2 boys are 0.2601 or 26.01%, of getting 2 girls are 0.2401 or 24.01% and of getting one boy and one girl are 0.4998 or 49.98%.

   The same may be extended to birth of *3 children*, possible combinations will be 3 boys (p^3), 3 girls (q^3), 1 boy and 2 girls ($3pq^2$), and 2 boys and 1 girl ($3p^2q$). To calculate the per cent chances of these combinations apply the formula:

   $(p + q)^3 = p^3 + q^3 + 3pq^2 + 3p^2q$
   $= 0.51^3 + 0.49^3 + 3 \times 0.51 \times 0.49^2 + 3 \times 0.51^2 \times 0.49$
   $= 13.27\% + 11.76\% + 36.74\% + 38.23\%$

2. In four ICDS projects, 20% of children under 6 years of age were found to be severely malnourished, i.e., in

grades III and IV. If only 4 children were selected at random from the four projects, what is the probability of 4, 3, 2, 1 and 0 (none) being severely malnourished? p = 20% = 0.2, q = (1 − 0.2) = 0.8 and n = 4.

As per terms of expansion of the binomial expression

$(p + q)^n = (p + q)^4 = p^4 + 4p^3q + 6p^2q^2 + 4pq^3 + q^4$

Substitute p and q values

$= (0.2)^4 + 4 \times (0.2)^3 \times 0.8 + 6 \times (0.2)^2 \times (0.8)^2 +$
$\quad 4 \times 0.2 \times (0.8)^3 + (0.8)^4$

$= 0.0016 + 0.0256 + 0.1536 + 0.4096 + 0.4096$

So probability of all 4 children being severely malnourished will be 0.16%, 3, 2, 1 and 0 (none) will be 2.56%, 15.36%, 40.96% and 40.96%, respectively.

Probability (Chances) from Shape of Normal Distribution or Normal Curve

If heights are normally distributed and total number of individuals such as 200 are taken as unity, then we know that 50% individuals are above the mean height and 50% below the mean height. The range, mean ± 1 SD covers 68% and mean ± 2 SD covers 95% of the individuals. So the probability of having height above mean + 2 SD = 2.5% and of having height below mean − 2 SD = 2.5%.

The probability of any observation or number of observations lying above or below, at any distance from the mean can be estimated (Chapter 5).

Similarly, the total area under the normal curve is taken as unity (one). For a normally distributed variable, the proportional area under any part of the curve will indicate the relative frequency or probability of observations between any two points on the horizontal scale. The probability or relative frequency of an observation falling in shaded areas (Fig. 7.1A) beyond mean, ± 2 SD is small, being only 2.5% + 2.5% = 5% or 0.05 out of 1.

Probability of an observation lying outside mean ± 3 SD will still be smaller, being 0.5% only (Fig. 7.1B). Z-test described in Chapter 5, gives the probability of location of

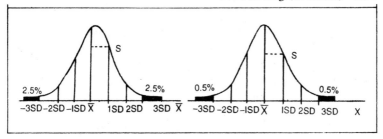

Fig. 7.1A and B: Probability or relative frequency of an observation falling beyond mean ± 2 SDs and ± 3 SDs, respectively

any observation in relation to mean in a normal frequency distribution.

Similarly, probability of sample values or results like means and proportions differing by chance from those of other samples or of population is determined from the shape of sampling distribution (Chapter 8).

If a sample is representative of a population, theoretically its mean should be equal to that of another representative sample or equal to population mean but it is not so. As per sampling distribution, 95% sample means lie within the limits of mean ± 1.96 standard error. Probability of values being higher and lower than this range is 5% (0.05 out of one). Conversely, probability of a mean not falling within the range—sample mean ± 2 SE is 95% (Fig. 7.2A). This applies to probability of differences between two sample means as well (Fig. 7.2B). The application of finding the relative frequency or probability of sample values will be better understood in subsequent chapters.

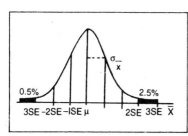

Fig. 7.2A: Normal distribution of means and probability of results falling beyond 2 SEs and 3 SEs

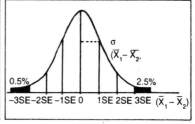

Fig. 7.2B: Normal distribution of differences between two means and probability of a difference falling beyond 2 and 3 SEs of difference

Probability of Calculated Values from Tables

Probability of calculated values occurring by chance in case of 't' and χ^2 is determined by referring to the respective tables. Probability or chances of dying or survival at any age are determined from the life tables constructed on the mortality experience of a large sample of population representative of both sexes and all ages. Modified life table methods are also used to find the chances of survival up to any point of time after a particular treatment or operation.

Chapter **8**

Sampling Variability and Significance

Measures of individual variability such as mean, standard deviation and shape of normal distribution or normal curve are dealt with in earlier chapters. In this Chapter we shall study the measures of group variability from sample to sample or sample to population. Such a study involves 3 steps.
1. *Selection of sufficiently large and random samples* representative of the population from which they are drawn. The techniques used are described in Chapter 6.
2. *Finding the probability or relative frequency of the sample* results, occurring by chance as explained in Chapter 7.
3. *Drawing the inference,* if the probability of a sample value is found higher or lower than the *probability of its occurrence* by chance as being discussed in this Chapter.

In individual variation, SD(s) and shape of normal distribution or curve were found to be good measures of dispersion of observations around the mean $(\overline{X})$. Now we shall determine if the standard error (standard deviation of means, $\overline{X}$) and shape of sampling distribution or normal curve are good measures of dispersion of means $(\overline{X})$ around the population mean (μ).

SAMPLING DISTRIBUTION

If successive samples from a population are drawn and their summary values like means and SDs are calculated, a series of different values are obtained (means, are $\overline{X}_1, \overline{X}_2, ..., \overline{X}_n$

and standard deviations SD_1, SD_2, ..., SD_n. They are variable because samples are variable but mean (μ) and SD (σ) of any particular population are constant. If the size of the samples is 30 or more, their summary values like means or SDs are close to those of the population.

Chance variation from sample to sample or from sample to universe is measured by **standard error** which is discussed in detail in the next Chapter.

If the number of samples is large, their values may be grouped as done in a frequency distribution table. It will be seen that the samples follow a normal distribution (Fig. 8.1). In any such sampling distribution:

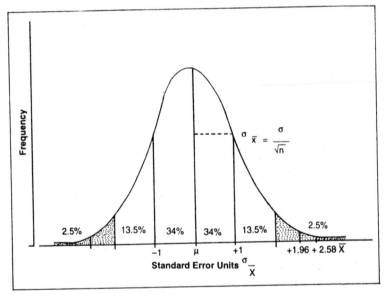

Fig. 8.1: Normal distribution of sample values (means)

i. Population value ± 1 SE limits include 68% of the sample values and fairly large number, i.e., 32% samples will have higher or lower values (16% on either side).

ii. Population value ± 1.96 SE limits include 95% of the sample values. Few, only 5% sample values, will fall

beyond this range or these limits. In other words, chances of such high or low values being normal will be 5%.
iii. Range defined by population value ± 2.58 SE includes 99% of the estimates, hence, an estimate higher or lower than that will be obtained by chance in 1% cases, i.e., very rarely. Such high or low value will probably be due to some factor(s) acting on the sample.

Thus 6 standard errors, three on either side of the universe value, cover almost the entire range of sample values.

If the number of samples is very large and class interval is made very small, the frequency polygon of sample values will give a smooth curve, i.e., a normal curve (Fig. 5.1) with characteristics already described in Chatper 5.

The normal distribution in samples is one of the standard distributions and it forms a basis for the tests of significance.

SIGNIFICANCE

After making experiments in medical problems, certain results like means and proportions are obtained which vary from sample to sample and sample to universe.

Measures of individual variability such as SD and shape of normal distribution as well as variability of sample values (means and proportions) such as SE and shape of sampling distribution have been explained.

Next is the stage of interpretation of results or drawing statistical inferences or conclusions. In other words, the observer or experimenter wants to know the significance of the difference he has observed in his result as compared with that of the population or with that of another worker, e.g., he finds that mean blood pressure of his sample is higher than that observed by another worker or another worker may find the cure rate with chloramphenicol to be higher than with tetracycline and so on.

The difference observed is expressed in terms of significance or probability or relative frequency of its occurrence by chance and is stated on the basis of sampling distribution.

118 Methods in Biostatistics

There are two basic methods of drawing the conclusion or knowing the significance of the results obtained.
1. The estimation of a population parameter from a sample statistic.
2. The testing of hypotheses about the population parameter.

Estimation of Population Parameter

We cannot draw large number of samples covering the entire population in order to find the population parameter (μ). So, we calculate the same from a sample statistic such as $\overline{X}$.

We then set up certain limits on both sides of the population mean (μ) on the basis of the fact that *means ($\overline{X}$) of samples of size 30 or more are normally distributed around the population mean (μ)*.

These limits are called the **confidence limits** and the range between the two is called the **confidence interval**. As per normal distribution of samples, we say with confidence or we are sure that 95% of the sample means ($\overline{X}$) will lie within the confidence limits of (Fig. 8.2) —

$\mu - 1.96$ SE ($^s\overline{X}$) and $\mu + 1.96$ SE ($^s\overline{X}$)

95% confidence interval thus obtained will contain 95% of sample means.

Conversely, *population mean (μ) will also fall within these confidence limits or lie in the confidence intervals* between (Fig. 8.3) —

$\overline{X} - 1.96$ SE ($^s\overline{X}$) and ($\overline{X} + 1.96$ SE ($^s\overline{X}$)

at 95% confidence interval. The range, thus obtained, will contain population mean in 95% cases. This also implies that any sample or universe value lying outside the range mean $\pm$ 1.96 SE will be rare. The probability or relative frequency of such occurrence by chance will be 5% or 0.05 out of one, i.e., once in 20 times.

If confidence limits are extended to cover wider interval or range between mean $\pm$ 2.58 SE, we can say with 99% confidence that other sample means as well as that of

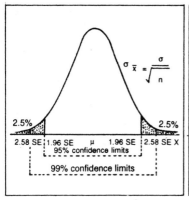 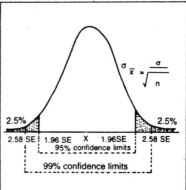

Fig. 8.2: Confidence limits of sample mean ($\bar{X}$) from the population mean (μ) are ($\mu \pm 1.96SE$ and $\mu \pm 2.58SE$)

Fig. 8.3: Confidene limits of population mean (μ) from the sample mean ($\bar{X}$) are ($\bar{X} \pm 1.96SE$ and $\bar{X} \pm 2.58SE$)

universe or population would fall in these limits. Any value lying outside this interval will be very rare. Its probability (p) will be 0.01 or 1%, i.e., once in 100 times.

Thus, confidence limits or confidence intervals help us to estimate the location of a sample mean ($\bar{X}$) if the population mean (μ) is known or vice versa to locate the population mean (μ) where sample mean ($\bar{X}$) is known.

We have found the *true value* or, *population* value from that of the sample but we cannot be confident or sure because we are dealing with a part of the population only howsoever big the sample may be. We would be wrong in 5% cases only if we place the population value within 95% confidence limits and in 1% cases only if we say the population mean lies within 99% confidence limits.

The limit of the region at which we no longer regard the chance to be operating is called the **level of significance**. It separates the shaded areas in one or two tails of the area under the normal curve form the plain area (Figs 8.4A, B and 8.5A, B).

If the chance limit is set at mean $\pm$ 1.96 SE, it implies 5% or 0.05 level of significance, also called the *critical level of significance* (Fig. 8.4A). A value lying beyond this area is said

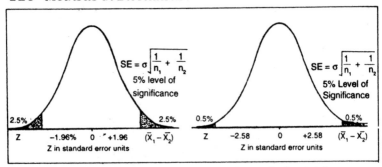

Figs 8.4A and B: Shaded areas indicate the levels of significance distributed in both the tails

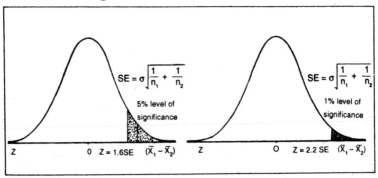

Figs 8.5A and B: Shaded areas indicate the level of significance lying at one end only

to be *significantly different* from the population value. As stated already, at this level, such extreme values will occur by chance only 5 times in 100 experiments.

If the line is drawn at a distance of 2.58 SE from the mean, its level of significance is said to be 1% (Fig. 8.4B). A value lying at this limit or beyond is *highly significant* because such a different value will be found by chance only once in 100 results (P, 0.01).

In most of the statistical studies, the levels of significance are set at 5% (P, 0.05), 1% (P, 0.01) and 0.5%. (P, 0.005). They are the yardsticks against which the probability (p) or relative frequency of our sample estimate is measured. Significant or insignificant indicates whether a value is likely to occur by chance or it is unlikely to occur by chance.

Testing of Statistical Hypothesis

To test statistical hypotheses about the population parameter or true value of universe, two hypotheses or presumptions are made to draw the inference from the sample value.

1. *A null hypothesis or hypothesis of no difference* (H_0) between statistic of a sample and parameter of population or between statistic of two samples. This hypothesis nullifies the claim that the experimental result is different from or better than the one observed already.

2. *The alternative hypothesis of significant difference* (H_1) stating that the sample result is different—greater or smaller than the hypothetical value of population, e.g. weight gain or loss due to new feeding regimen.

By this we shall adopt a procedure to choose between null hypothesis (H_0) and alternate hypothesis (H_1) by applying relevant statistical technique.

A test of significance such as Z-test is performed to accept the null hypothesis H_0 or to reject it and accept the alternative hypothesis H_1.

To make minimum error in rejection or acceptance of H_0, we divide the sampling distribution or the area under the normal curve into two regions or zones (Fig. 8.6).

   i. A zone of acceptance
   ii. A zone of rejection

i. *Zone of acceptance* If the result of a sample falls in the plain area, i.e., within the mean ± 1.96 SE the null hypothesis is accepted, hence this area is called the *zone of acceptance* for null hypothesis.

ii. *Zone of rejection* If the result of a sample falls in the shaded area, i.e., beyond mean ± 1.96 SE it is siginificantly different from the universe value. Hence, the H_0 of no difference is rejected and the alternate H_1 is accepted. This shaded area, therefore, is called the zone of rejection for null hypothesis. It may be distributed at both ends (Fig. 8.6) or lie at one end of the area under the normal curve (Figs 8.7 and 8.8).

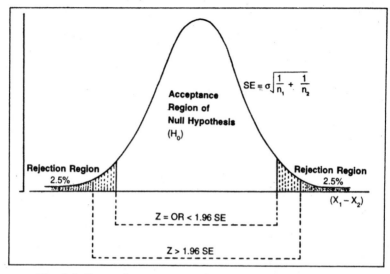

Fig. 8.6: Zones of acceptance and rejection of null hypothesis. Rejection region distributed in two tails

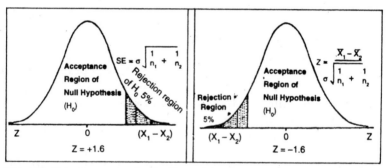

Fig. 8.7: Rejection region for the H_0 distributed in right tail (higher extremes)

Fig. 8.8: Rejection region for H_0 distributed in left tail (lower extremes)

Type I and Type II Errors

As a routine we follow these zones of acceptance and rejection to interpret the significance of result or estimated difference from the universe parameters at 5% level of significance.

In certain circumstances, the null hypothesis of no difference is rejected even when the estimate falls in the zone

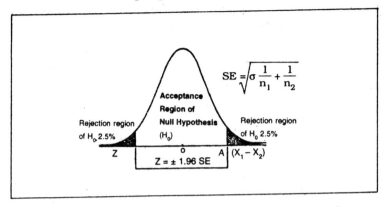

Fig. 8.9: Committing Type I error, rejecting H_0 at the point A in the zone of acceptance

of acceptance at 5% level say at point A (Fig. 8.9). It means we are changing the level of significance from 5% to 6, 8 or 10%, etc. This is committing Type I error also called 'α' by changing the level of significance. The extent to which H_0 may be rejected depends on the investigator and the circumstances such as trial of two drugs when he may think that the difference at 10% level of significance is enough.

There are other situations when H_0 is accepted when it should have been rejected because the estimate falls in the zone of rejection, i.e., in shaded areas, say at point B, (Fig. 8.10). Here, we are changing the level of acceptance from 5% to 4, 3, 2 or 1% level of significance. This is committing Type II error also called 'β'. By this we make the test more stringent, e.g., while comparing heights or weights

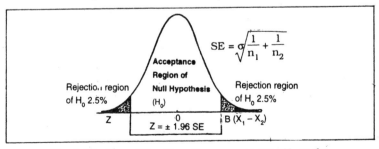

Fig. 8.10: Committing Type II error, accepting H_0 at the point B in the zone of rejection

of Punjabis and Bengalis, though the difference in terms of SE falls in the shaded area at a significant level of 4%, we may still accept H_0.

In such cases we increase the size of sample and confirm the inference. So, Type II error is not so serious because it only needs confirmation of result by changing the level of significance.

In the one-tailed test, if the result is interpreted at 10% (equivalent to 5% in two-tailed test) level, the risk of committing Type I error is reduced.

It is desirable to interpret the results at 3% level in place of 1% or 5% rather than committing Type II error.

To **summarise** when a statistical hypothesis is treated there are 4 possible ways of interpreting the result:
1. The hypothesis H_0 is **true** and our test accepts it because the result falls within the zone of acceptance at 5% level.
2. The hypothesis H_0 is **false** and our test rejects it because the estimate falls in shaded area of rejection.

These two interpretations are a routine but the other two interpretations may lead to errors—
3. Hypothesis H_0 is **true** still **it is rejected,** though the estimate falls in acceptance zone at 5% level (Fig. 8.9) in plain area (Type I error).
4. The hypothesis H_0 is **false** but **it is accepted**, though the estimate falls (Fig. 8.10) in the zone of rejection (Type II error).

Inference	Accept it	Reject it
Hypothesis is true	Correct decision	Type I error
Hypothesis is false	Type II error	Correct decision

Type I error is usually fixed in advance by choice of the level of significance employed in the test, e.g., if it is fixed at 5% level, we know that we shall be rejecting on an average, 5% of all the true hypotheses. It may be noted that Type I error can be made as small as desired by changing the level of significance.

In medical studies, Type I error is more serious as compared to Type II. The level of significance is usually set at 5%, 1% or 0.5% (Figs 8.4A, B and 8.5A, B) and the rejection region (Figs 8.6–8.8) is distributed in either one or both the tails of the area under normal curve. To minimise the error of chance, we should take as large a random sample as possible and interpret the results at 5%, i.e., *critical level of significance*. If the sample is not large enough, draw the inference at 5% level and confirm it by increasing the size of the sample.

The probability values (P) for one- and two-tailed tests are given separately in the probability tables. Compare your result at P, 0.10 for 5% level of significance instead of at P, 0.05, in one-tailed tests.

TESTS OF SIGNIFICANCE

These tests are mathematical methods by which the probability (p) or relative frequency of an observed difference, occurring by chance is found. It may be a difference between means or proportions of sample and universe or between the estimates of experiment and control groups.

Methods of determining the significance of difference are discussed to draw inferences and conclusions. Common tests in use are, 'Z' test, 't' test and 'χ^2' test. The first two tests express the difference observed in terms of standard error (SE) which is a measure of variation in sample estimates that occur by chance. The value of ratio between observed difference and SE, is that of the table, which gives the highest obtainable values compared with difference occurring by chance at different levels of significance (P, 0.10, 0.05, 0.01, etc.).

The stages in performing a test of significance are:
1. State the null hypothesis of no or chance difference and the alternative, e.g., vitamins A and D make no difference in growth or alternatively they play a positive or significant role in promoting growth.

2. Determine P, i.e., probability of occurrence of your estimate by chance, i.e., accept or reject the null hypothesis.
3. Draw conclusion on the basis of P value, i.e., decide whether the difference observed is due to chance or play of some external factors on the sample under study.

Z-test is described here while other tests and measures of variation in sample estimates will be explained in subsequent Chapters.

Z-test

Recapitulate normal deviate 'Z' (Chapter 5) discussed under normal distribution of a variable within one sample. There, the difference (Z) from $(\bar{X})$ or (μ) was measured in terms of SDs (s) or (σ).

If the samples are large in size (more than 30), the means of samples in a population of differences between pairs of samples in a population also follow normal distribution.

When Z-test is applied to the sampling variability, the difference observed between a sample estimate and that of population is expressed in terms of SE instead of SD. The score of value of the ratio between the observed difference and SE is called 'Z'.

If the distance in terms of SE or Z score falls within mean ± 1.96 SE, i.e., in the zone of acceptance (95% confidence limits) the H_0 is accepted. The distance from the mean at which H_0 is rejected is called the **level of significance.** It falls in the zone of rejection for H_0, shaded areas under the curves and it is denoted by letter P which, indicates the probability or relative frequency of occurrence of the difference by chance. Greater the Z value, lesser will be the P. As per normal or sampling distribution, only 5% of Z values occur by chance beyond mean ± 1.96 SE, P at 5% level is written as 0.05% and at 2.5% level as 0.025 and so on (Figs 8.1–8.3). Z value at the significant level will be significantly different, higher or lower, than the hypothetical or theoretical (population) value.

Sampling Variability and Significance

Thus, the Z-test for means has two applications:
1. To test the significance of difference between a sample mean ($\bar{X}$) and a known value of population (μ).
   Observed difference between samples

$$Z = \frac{\text{Mean } (\bar{X}) - \text{population mean } (\mu)}{\text{SE of sample mean}}$$

2. To test the significance of difference between two sample means or between experiment sample mean and a control sample mean.

$$Z = \frac{\text{Observed difference between two sample means}}{\text{SE of difference between two sample means}}$$

The four prerequisites to apply Z-test for mean are:
i. The sample or samples must be randomly selected.
ii. The data must be quantitative.
iii. The variable is assumed to follow normal distribution in the population.
iv. The sample size must be larger than 30.

If SD of populations is known, Z-test can still be applied even if the sample is smaller than 30.

To determine the significance of Z value, the probability (P) value is found from the following table, constructed on the basis of normal distribution.

Z:	1.6	2.0	2.3	2.6
	(1.65)	(1.96)	(2.2)	(2.58)
P:	0.10	0.05	0.02	0.01

If the Z value increases, the P value or probability of an event happening by chance decreases and alternative happening due to some external factor is to be considered.

NB — Figures in brackets give more precise values of Z. In practice, values of Z without brackets are followed.

One-tailed and Two-tailed Z-tests

Z values on each side of mean are calculated as + Z or as – Z. A result larger than ($\bar{X}$) will give + Z and the result smaller than ($\bar{X}$) will give – Z.

If IQ of an individual is 70 and the mean IQ is 100 with a SD of 15.

$$Z = \frac{X - \overline{X}}{15} = \frac{70 - 100}{15} = -2 \text{ (P is } < 0.5)$$

If IQ of another individual is 115.

$$Z = \frac{115 - 100}{15} = +1$$

In a normal distribution, the highest and the lowest values are at the two extreme ends or tails of the area under a normal curve of a frequency distribution. Relative frequency or probability (P) of any result occurring by chance at two tails beyond $(\overline{X}) \pm 1.96$ SE will be 5% (Figs 8.2, 8.3 and 8.6) or 0.05 (total at both ends) and 2.5% or 0.025 at each end or tail (Figs 8.5A and B, 8.7 and 8.8).

In a test of significance, when one wants to determine whether the mean IQ of malnourished children is different from that of well nourished and does not specify higher or lower, the P value of an experiment group includes both sides of extreme results at both ends of scale, and the test is called **two-tailed test** (Figs 8.4A and B and 8.6). In case of 5% level of significance, probability (P) will be 2.5% (0.025 at each end). So the result is compared with table value at the probability level of 0.05 taking into account the probability of such difference occurring by chance at both ends or tails (0.025 + 0.025 = 0.05). The difference is being tested for significance but the direction is not specified.

If one wants to know specifically whether a result is larger or smaller than what can occur by chance (plus or minus 1.96 SE), he excludes one end of the scale on theoretical ground. He specifies the direction on plus or minus side. The significant level or P value will apply to relative end only, e.g., when we want to know if the malnourished have lesser mean IQ than the well nourished or English boys are taller than Indian boys, the result will lie at one end or tail of the distribution (Figs 8.5A and B, 8.7 and 8.8). So when the result of an experiment is compared with table value at P = 0.05, the probability of higher or lower result occurring by chance will be half of 0.05, i.e., 0.025. Such a test of significance is called **one-tailed test**. If P value is obtained as 0.02 in one-tailed test, the significance level will be half, i.e., 0.01.

If we want to know whether one particular drug is *better* than the other, it will be *one-tailed* test. When you want to know the action of a particular drug is *different* from that of another, it will be *two-tailed test*.

The probability values for Z, t and χ^2 are given in the tables for both the tails together in the Appendices at the end for P—0.10, 0.05, 0.02, etc. To assess the significance of a result obtained in Z- or other tests of significance, refer the value to that given under appropriate P value for two-tailed test.

In case of one-tailed test, halve the value of P but it will be simpler or easier to remember that in one-tailed test, we refer the result against value under P = 0.10, for 5% level of significance. If it is greater, the probability of result occurring by chance will be < 0.05, say result is significant at 5% level of significance. If it is smaller, the probability will be greater than 0.05, so insignificant at 5% level.

In some books on statistics, the probabilities for one- and two-tailed tests are given separately.

Chapter **9**

Significance of Difference in Means

Tests of significance of difference in means are discussed under two heads.
1. Z-test for large samples.
2. t-test for small samples applied as—
    i. Unpaired t-test (two independent samples); and
    ii. Paired t-test (single sample correlated observations).

The two essential conditions for application of these tests are:
   i. Samples are selected randomly from the corresponding populations.
   ii. There should be homogeneity of variances in the two samples.

First condition could be taken care of by adopting appropriate sampling procedure such as simple random sampling.

To test the *homogeneity* of variances, Fisher's F-test also called **variance ratio test** is applied (described later in the chapter). If the difference is found to be insignificant, there is homogeneity of variances and the Z-test and or t-test can be carried out. If the ratio is significant, i.e., the samples are heterogeneous, a different test in place of t-test has to be applied, for which reader may refer to books mentioned under Bibliography.

For application of either of the two tests, 'Z' or 't', standard error of mean, i.e., a unit that measures chance variation has to be understood.

STANDARD ERROR OF MEAN (SE $\overline{X}$)

Whatever be the sampling procedure or the care taken while selecting the sample, the sample estimates of statistics, ($\overline{X}$, s or p) will differ from population parameters (μ, σ or P) because of chance or biological variability. Such a difference between sample and population values is measured by statistic known as sampling error or **standard error** ($^s\overline{X}$) or (SEp).

Standard error is thus a measure of chance variation and it does not mean error or mistake.

Calculation of Standard Error of Mean ($^s\overline{X}$)

SE or SD of means of samples drawn from a particular population could be calculated if means of large number of samples were known. Follow the same procedure as was done to calculate SD of one series. To calculate the SE, find the mean (μ) of the sample means ($\overline{X}$) and then the differences of individual means from this grand mean ($\overline{X} - \mu$). Standard deviation of a series of means or

$$^s\overline{X} = \sqrt{\frac{\Sigma(\overline{X} - \mu)^2}{n - 1}}$$

In practice, it is not possible to take large number of samples, find their means $\overline{X}_1, \overline{X}_2, ..., \overline{X}_n$ to work out population mean and then calculate the standard error by this formula. Usually only one large sample is drawn and its standard deviation is calculated. Then SE of mean is calculated by the following formula.

$$SE\ \overline{X}\ or\ (^s\overline{X}) = \frac{SD}{\sqrt{n}} = \frac{s}{\sqrt{n}}$$

Thus, SE of mean is SD of the sample divided by the square root of the number of observations in the sample. Its value varies directly with the size of SD. Greater the SD greater will be the SE as will happen in a small sample. Sampling error or chance variation has to be minimised by reducing the standard deviation which can be done only by taking a large sample. An individual unusually tall, say of height 190 cm, would raise the mean and standard deviation of height, markedly in a sample of 10 but would have little influence in a large sample of 100. The standard error varies inversely with the square root of the number. If the sample is increased 100 times, the SE is reduced to almost one-tenth.

Applications and Uses of the SE of Mean in Large Samples

1. To work out the *limits of desired confidence* within which the population mean, µ, would lie. As per sampling distribution, the range of these limits would be sample mean ± 1.96 SE in 95 estimates out of 100. The odds of population mean lying outside these limits will be 1:19. The range of these limits will be $\overline{X} \pm 2 = \frac{s}{\sqrt{n}}$ (Fig. 8.1).

2. To determine *whether the sample is drawn from a known population or not* when its mean (µ) is known. If the sample mean is larger than the known population mean ± 1.96 SE, 95% chances are that the sample is not drawn from the same population or the sample is under the influence of some external factor. We may be wrong in 5 cases out of 100 at 95% confidence limits (Fig. 8.2).

Find the probability of the difference observed by Z-test.

a. $Z = \frac{\overline{X} - \mu}{\sigma/\sqrt{n}}$ when µ and σ are known.

b. $Z = \frac{\overline{X} - \mu}{\sigma/\sqrt{n}}$ when µ is known but σ is not known.

To determine significance of the value refer to table on page 127 of chapter 8 (Z and P value).

3. To find the SE of difference between two means to know if the observed difference between the means of two samples is real and statistically significant or it is apparent and insignificant due to chance (Figs 8.4 A,B and 8.5 A,B).

Significance of Difference in Means

4. To calculate the *size of sample* in order to have desired confidence limits, if SD of population is known, as explained under sampling in Chapter 6.

Examples

NB—For ease in calculation, 2 SE are taken instead of 1.96 SE for 95% confidence limits in all the examples that follow. This change is not material. It raises the confidence limits to 95.45% and makes the test a little more *strict*.

1. Systolic blood pressure of 566 males was taken. Mean BP was found to be 128 mm and SD 13.05 mm. Find 95% confidence limits of BP within which the population mean would lie.

$$s\overline{X} = \frac{s}{\sqrt{n}} = \frac{13.05}{\sqrt{566}} = 0.55$$

Mean BP + 2 SE = 128.8 + 2 × 0.55 = 129.9

Mean BP − 2 SE = 128.8 − 2 × 0.55 = 127.7

Confidence limits for population mean will be in the range of 127.7 and 129.9 in 95% of cases. If mean BP of another sample is found to be beyond 95% confidence limits, the sample is unlikely to have been drawn from the same universe or it may be under the influence of some external factor or cause such as tension, excitement, drug, etc. We could be wrong in 5 cases out of 100.

2. Index of brightness of 50 boys and 50 girls gave following values:

	Mean	SE
Boys	91.2	5.23
Girls	90.8	4.41

Find 95% confidence limits of mean index of brightness in population of both.

At 95% confidence limits of mean, index of brightness for all boys

= 91.2 ± 2(5.23), i.e., 80.74 to 101.66 units

At 95% confidence limits of mean, index of brightness for all girls

= 90.8 ± 2(4.41), i.e., 81.98 to 99.62

3. Height of 50 boys and 50 girls gave following values:

	Mean X	SD s	SE $s\bar{X}$
Girls	147.4 cm	6.6 cm	0.93 cm
Boys	151.6 cm	6.3 cm	0.89 cm

Find the 95% confidence limits within which the height of all girls and boys would lie.

Limits of mean height for all girls
$= 147.4 \pm 2 \times 0.93 = 145.54$ to 149.26 cm

Limits of mean height for all boys
$= 151.6 \pm 2 \times 0.89 = 149.82$ to 153.38 cm

Any mean index of brightness or height mean above and below these limits will indicate that sample is not drawn from the same universe at 95% limits of confidence or it may be under the play of some influence or factors such as nutritional status.

4. SD of blood sugar level in a population is 6 mg%. If population mean is not known, within what limits is it likely to lie if a random sample of 100 has a mean of 80 mg%?

$$SE = \frac{6}{\sqrt{100}} = \frac{6}{10} = 0.6$$

At 95% confidence limits, the population mean would lie in the range of 78.8 to 81.2 mg.

5. In a population sample of children with mean height = 66 cm and SD = 2.7 cm, can a sample of 100 with a mean height of 67 cm occur easily? If you find that the probability is low ($P < 0.01$), what does it indicate?

$$^s\bar{X} = \frac{s}{\sqrt{n}} = \frac{2.7}{\sqrt{100}} = \frac{2.7}{10} = 0.27$$

Since 67 is more than $66 + 3\,^s\bar{X}$, i.e., $66 + 3 \times 0.27 = 66.81$ cm, this sample cannot easily occur in this population. P is less than 0.01. It indicates that 99% chances are that the sample is not drawn from the same population, may be the children in the sample belong to upper age group or class.

Probability (p) of its being drawn from the same universe is less than 1% (or $p < 0.01$).

STANDARD ERROR OF DIFFERENCE BETWEEN TWO MEANS OF LARGE SAMPLES

SE $(\overline{X}_1 - \overline{X}_2)$ or $^s(\overline{X}_1 - \overline{X}_2)$

If independent, large and random samples are drawn in pairs, repeatedly from the same population and each time the difference between the two means of each pair is calculated, there will be a population of differences or a series of differences of the paired samples as below:

$$(\overline{X}_{1a} - \overline{X}_{1b}), (\overline{X}_{2a} - \overline{X}_{2b}), ..., (\overline{X}_{na} - \overline{X}_{nb})$$

The means of these differences from the population will be zero (0) from which the differences in pairs of means deviate to plus or minus side of 0. They also follow normal distribution around the mean 0 just as means of samples $(\overline{X})$ do around the population mean (μ).

Frequency distribution of the differences gives a normal curve. The standard deviation of such a distribution of differences is known as **standard error of difference between two means.**

In practice, it is not possible to find large number of differences in pairs of sample means and then find the SE of these differences. The test is applied to one pair directly if standard deviation of means of two samples are known. Assumption of no difference between the two sample means is made as per null hypothesis (Fig. 8.6).

According to the null hypothesis, if the samples drawn are random and of sufficiently large size, their means should not differ from the population mean (μ) or from means of each other. The difference should be nil or zero (Fig. 8.4 A,B). Non-zero values of differences should be insignificant, if they are not larger or smaller than 1.96 times the SE of difference in 95% cases. Values larger or smaller than 3 times the standard error will be very rare, less than once in 100 cases.

If the observed difference between the two means is greater than 1.96 times the standard error of difference, it is significant at 5% level of significance (Fig. 8.4A). The difference is unusual and may be due to influence of some external factor.

If the observed difference is greater than 3 times the SE, it is real variability in more than 99% cases, and biological or due to chance in less than 1% cases (Fig. 8.4B).

Apply normal deviate (Z) test as explained in Chapter 8 for large samples.

Calculation of Standard Error of Difference

SE of difference is denoted as $^s(\overline{X}_1 - \overline{X}_2)$ or $\sigma(\overline{X}_1 - \overline{X}_2)$. Following formulae are applied for its calculation:

1. SE of difference is the square root of the sum of squares of two standard errors of means, so

$$^s(\overline{X}_1 - \overline{X}_2) = \sqrt{\left(\frac{SD_1}{\sqrt{n_1}}\right)^2 + \left(\frac{SD_2}{\sqrt{n_2}}\right)^2}$$

$$= \sqrt{\frac{SD_1^2}{n_1} + \frac{SD_2^2}{n_2}} \text{ or } \sqrt{\frac{s_1^2}{n_1} + \frac{s_2^2}{n_2}}$$

This formula is applied when μ and σ of population are not known but the samples are pretty large. Thus, the use of variance of two samples is made in this formula.

$$Z = \frac{\overline{X}_1 - \overline{X}_2}{^s(\overline{X}_1 - \overline{X}_2)}$$

2. $S(\overline{X}_1 - \overline{X}_2) = \sigma\sqrt{\frac{1}{n_1} + \frac{1}{n_2}}$

This is applied when σ is known but μ is not known.

$$Z = \frac{(\overline{X}_1 - \overline{X}_2)}{\sqrt{\sigma\left(\frac{1}{n_1} + \frac{1}{n_2}\right)}}$$

3. $SE = \sqrt{\frac{SD^2}{n_1} + \frac{SD^2}{n_2}} = \sqrt{SD^2\left(\frac{1}{n_1} + \frac{1}{n_2}\right)}$

$SE = SD \times \sqrt{\left(\frac{1}{n_1} + \frac{1}{n_2}\right)}$ because $SD = \sqrt{SD^2}$

This formula is applied most often when the population variance is not known. So instead of that we find the *combined variance of both the samples* as given below.

The square root of combined variance will give the SE.

$$\frac{\text{Sum of squares in one sample}}{n_1 - 1} + \frac{\text{Sum of squares in the other}}{n_2 - 1}$$

$$= \frac{\Sigma(\overline{X}_1 - \overline{X}_1)^2 + \Sigma(\overline{X}_2 - \overline{X}_2)^2}{n_1 + n_2 - 2} = s^2$$

The square root of combined variance will give the SE.

$$\text{SE or } \frac{\overline{X}_1 - \overline{X}_2}{s\sqrt{\frac{1}{n_1} + \frac{1}{n_2}}}$$

$$Z = \frac{(\overline{X}_1 - \overline{X}_2)}{s\sqrt{\frac{1}{n_1} + \frac{1}{n_2}}}$$

Application of SE of Differences

The test, $^s(\overline{X}_1 - \overline{X}_2)$ is required to be applied very often in medical practice. Means of a normally distributed variable in the two like or unlike group are compared such as of height, weight, BP, pulse rate, etc. The action of a drug on a variable such as BP or pulse rate is compared in two groups when a placebo is given to the control group. Same way, the action of two different drugs or of two different doses of the same drug can be compared and so on. As this test is based on normal distribution, the samples should sufficiently be large and random.

Start with the assumption of no difference or insignificant difference, i.e., null hypothesis, find the value of Z, i.e., the ratio of observed difference to SE of difference as stated above in all the three methods of calculation of standard error of difference.

$$Z = \frac{\overline{X}_1 - \overline{X}_2}{{}^s(\overline{X}_1 - \overline{X}_2)}$$

Refer the Z value to find the probability of the observed difference occurring by chance, as per Table 8.1.

Examples

1. In a study on growth of children, one group of 100 children, had a mean height of 60 cm and SD of 2.5 cm while another group of 150 children had a mean height of 62 cm and SD of 3 cm. Is the difference between the two groups statistically significant?

Since µ and σ are not known and samples are much larger than 30, apply formula No. 1.

SE of difference of SE $(\overline{X}_1 - \overline{X}_2)$

$$= \sqrt{\frac{SD_1^2}{n_1} + \frac{SD_2^2}{n_2}} = \sqrt{\left(\frac{(2.5)^2}{100} + \frac{(3)^2}{150}\right)}$$

$$= \sqrt{\frac{6.25}{100} + \frac{9}{150}} = \sqrt{0.1225} = 0.35$$

$$Z = \frac{\overline{X}_1 - \overline{X}_2}{{}^s(\overline{X}_1 - \overline{X}_2)} = \frac{60 - 62}{0.35} = -5.71 \qquad ...(A)$$

Alternatively, if formula No. 3 is applied, combined variance of two samples can be found as below:

$$s_1^2 = \frac{\Sigma(X_1 - \overline{X}_1)^2}{n_1 - 1} \text{ or } \Sigma(X_1 - \overline{X}_1)^2 = s_1^2 \times (n_1 - 1)$$

$$s_2^2 = \frac{\Sigma(X_2 - \overline{X}_2)^2}{n_2 - 1} \text{ or } \Sigma(X_2 - \overline{X}_2)^2 = s_2^2 \times (n_2 - 1)$$

Both s and n are given so
In samples one and two

$$\Sigma(X_1 - \overline{X}_1)^2 = (2.5)^2 \times (100 - 1) = 6.25 \times 99 = 618.75$$

$$\Sigma(X_2 - \overline{X}_2)^2 = (3.0)^2 \times (150 - 1) = 9 \times 149 = 1341.00$$

Combined variance

$$= \frac{\text{Sum of squares of both the samples}}{n_1 + n_2 - 2}$$

$$= \frac{618.75 + 1341}{99 + 149} = \frac{1959.75}{248}$$

Combined SD $= \sqrt{\frac{1959.75}{248}} = 2.81$

SE of difference =

$${}^s(\overline{X}_1 - \overline{X}_2) = SD \times \sqrt{\frac{1}{n_1} + \frac{1}{n_2}} = 2.81 \sqrt{\frac{1}{100} + \frac{1}{150}}$$

$$= 2.81 \sqrt{\frac{3+2}{300}} = 2.81 \sqrt{\frac{1}{60}}$$

$$= 2.81 \times \sqrt{0.0166} = 2.81 \times 0.13 = 0.3628$$

$$Z = \frac{\text{Observed difference}}{SE} = \frac{60 - 62}{0.3653} = -5.47 \qquad ...(B)$$

(A) and (B) Z values are almost the same.

The observed difference calculated by both the methods is more than 3 times the SE hence highly significant. The growth is more in the second group than in the first.

2. In a nutritional study, 100 children were given a usual diet and vitamins A and D tablets. After 6 months, their average weight was 30 kg with SD of 2 kg while the average weight of the second comparable group of 100 children who were taking the usual diet only was 29 kg with SD of 1.8 kg. Can we say that vitamins A and D were responsible for this difference?

Calculate the SE and find if the difference is significant. If it is so, the vitamins are responsible. As per formula No. 1.

$$SE = \sqrt{\frac{(2)^2}{100} + \frac{(1.8)^2}{100}} = \sqrt{\frac{4 + 3.24}{100}} = \sqrt{0.0724} = 0.27$$

The ratio of observed difference to SE

$$Z = \frac{30 - 29}{0.27} = \frac{1}{0.27} = 3.7$$

By formula No. 3 also it comes to 3.7

As the value of the ratio (Z) is more than three times the SE, the observed difference is highly significant. Thus, the vitamins played a role in weight gain.

3. Find the significance of difference in the mean heights of 50 girls and 50 boys with following values:

	Mean	SD	SE
Girls	147.4 cm	6.6 cm	0.93 cm
Boys	151.6 cm	6.3 cm	0.89 cm

SE of difference between two means or

$$SE = \sqrt{\frac{SD_1^2}{n_1} + \frac{SD_2^2}{n_2}}$$

$$= \sqrt{\frac{(6.6)^2}{50} + \frac{(6.3)^2}{50}} = \sqrt{\frac{43.56}{50} + \frac{39.69}{50}}$$

$$= \sqrt{\frac{83.25}{50}} = \sqrt{1.665} = 1.29$$

$$Z = \frac{\text{Observed difference}}{SE} = \frac{151.6 - 147.4}{1.29}$$

$$= \frac{4.2}{1.29} = 3.26$$

By formula 3 also Z = 3.25.

The observed difference, 4.2 cm is more than 3 times the SE so it is significant at 99% confidence limits. The boys are taller than the girls.

4. Find the significance of difference in the mean index of brightness of 50 boys and 50 girls with the following values.

	Mean	SD	SE
Boys	91.2	37.0	5.23
Girls	90.1	31.2	4.41

$$SE = \sqrt{\frac{(SD_1)^2}{n_1} + \frac{(SD_2)^2}{n_2}} = \sqrt{\frac{(37)^2}{50} + \frac{(31.2)^2}{50}} = 6.8$$

$$Z = \frac{\overline{X}_1 - \overline{X}_2}{SE} = \frac{91.2 - 90.1}{6.8} = \frac{1.1}{6.8} = 0.16$$

By formula 3 also Z = 0.16.

Being much less than 1.96 at 95% confidence limits, the difference is probably due to chance, hence not signifcant.

NB—To calculate $^s(\bar{X}_1 - \bar{X}_2)$ any of the three formulae can be applied if the size of both the samples is larger than 30.

SIGNIFICANCE OF DIFFERENCE BETWEEN MEANS OF SMALL SAMPLES BY STUDENT'S t-TEST

Small samples or their Z values do not follow normal distribution as the large ones do. So, the Z value based on normal distribution will not give the correct level of significance or probability of a small sample value occurring by chance. In case of samll samples, t-test is applied instead of Z-test. It was designed by W.S. Gossett whose pen name was Student. Hence, this test is also called **Student's t-test**.

The ratio of observed difference between two means of small samples to the SE of difference in the same is denoted by letter 't'. Gossett showed that the ratio follows different distribution called the 't' distribution. This 't' corresponds to Z in large samples but the probability of occurrence 'p' of this calculated value is determined by reference to 't' table.

Fisher's table (Appendix II) gives the highest obtained values of 't' under different probabilities 'P' in decimal fractions—0.10, 0.05, 0.01 and 0.001 corresponding to the degrees of freedom (df), serially numbered.

Probability 'p' of occurrence of any calculated value of 't' is determined by comparing it with the value given in the row of table corresponding to the df derived from the number of observations in the samples under study.

Probability converted into percentage is stated as level of significance. $P = 0.05$ may be stated as significant at 5% level (equivalent to 95% confidence limits in Z-test).

If the calculated 't' value exceeds the value given under $P = 0.05$ in the table, it is said to be significant at 5% level and null hypothesis (H_0) is rejected and alternate hypothesis (H_1) is accepted.

Degrees of Freedom (df)

The quantity in the denominator which is one less than the independent number of observations in a sample is called **degrees of freedom** and used in preference to sample size. In unpaired t-test of difference between two means,

= $n_1 + n_2 - 2$ where n_1 and n_2 are the number of observations in each of the two series.

In paired t-test df = $n - 1$.

Application of t-test

It is applied to find the significance of difference between two means as:
1. Unpaired t-test, and as
2. Paired t-test.

Criteria for Applying t-test

1. Random samples
2. Quantitative data
3. Variable normally distributed
4. Sample size less than 30.

Criteria differ from Z test in condition No. 4.

Unpaired t-test

This test is applied to unpaired data of independent observations made on individuals of two different or separate groups or *samples drawn from two populations*, to test if the difference between the two means is real or it can be attributed to sampling variability such as between means of the control and experimental groups.

As per null hypothesis (H_0), it is assumed that there is no real difference between the means of two samples, if the samples are taken at random and drawn independently from the same population. Following steps are taken to test the significance of difference.
1. Find the observed difference between means of two samples ($\overline{X}_1 - \overline{X}_2$).
2. Calculate the SE of difference between the two means, i.e., ($\overline{X}_1 - \overline{X}_2$).

This measure of variation in a variable will determine the limits of chance or biological variation.

3. Calculate the 't' value, i.e., the ratio between the observed difference and its SE by substituting the above values in the formula,

$$'t' = \frac{\overline{X}_1 - \overline{X}_2}{SE} \text{ and } SE = \sigma\sqrt{\frac{1}{n_1} + \frac{1}{n_2}}$$ if 't' is not known, use <u>combined</u> variance, but cannot apply the formula. $SE = \sqrt{\frac{s_2^2}{n_1} + \frac{s_2^2}{n_2}}$ as in Z-test.

4. Determine the pooled degrees of freedom from the formula
    $df = (n_1 - 1) + (n_2 - 1) = n_1 + n_2 - 2$.
5. Compare calculated value with the table value (Appendix II) at particular degrees of freedom to find the level of significance in two-tailed test. In one-tailed test, compare your result with values given under P = 0.10 and P = 0.02. If it is higher, it is significant at 5% level (P = 0.05) and 1% level (P = 0.01), respectively, i.e., significance level has to be halved when 't' table does not give the probabilities separately for one-and two-tailed tests as explained under Z-test.

Examples

1. In a nutritional study, 13 children were given a usual diet plus vitamins A and D tablets while the second comparable group of 12 children was taking the usual diet. After 12 months, the gain in weight in pounds was noted as given in the table below. Can we say that vitamins A and D were responsible for this difference?

Children on vitamins (Group A)		Children on usual diet (Group B)	
X	X^2	X	X^2
5	25	1	1
3	9	3	9
4	16	2	4
3	9	4	16
2	4	2	4
6	36	1	1
3	9	3	9
2	4	4	16
3	9	3	9
6	36	2	4
7	49	2	4
5	25	3	9
3	9	–	–
52	240	30	86

	Group A	Group B
n	13	12
$\overline{X}$	4	2.5
ΣX^2	240	86
$\dfrac{(\Sigma X)^2}{n}$	$\dfrac{(52)^2}{13} = 208$	$\dfrac{(30)^2}{12} = 75$

$$\Sigma(X - \overline{X})^2 = \Sigma X^2 - \frac{(\Sigma X)^2}{n}$$

$$= 240 - 208 = 32 \qquad 86 - 75 = 11$$

SD^2 or combined variance

$$= \frac{\Sigma(X - \overline{X})^2 \text{ of group A} + \Sigma(X - \overline{X})^2 \text{ of group B}}{\text{Total number in two groups} - 2}$$

$$= \frac{32 + 11}{13 + 12 - 2} = \frac{43}{23} = 1.87$$

$SD = \sqrt{1.87} = 1.37$

$$SE = SD \times \sqrt{\left(\frac{1}{13} + \frac{1}{12}\right)}$$

$$= 1.37 \times \sqrt{0.16} = 1.37 \times 0.4 = 0.548$$

$${}^*t_{23} = \frac{\overline{X} \text{ of group A} - \overline{X} \text{ of group B}}{SE}$$

$$= \frac{4 - 2.5}{0.548} = \frac{1.5}{0.548} = 2.74$$

**Pooled degrees of freedom = $n_1 + n_2 - 2 = 13 + 12 - 2 = 23$.

At 23 df the highest obtainable value of 't' at 5% level of significance is 2.069 as found on reference to 't' table (Appendix II).

The 't' value in this experiment is calculated at 2.74 which is much higher than the highest 2.069 obtainable by chance. Thus, the probability of occurrence (P) of the value obtained (2.74) by chance is much less than 0.05, the critical or 5%

---

*t_{23} means t value at 23 degrees of freedom
**Unpaired t-test is also called pooled t-test when sum of squares and degrees of freedom are pooled.

level of significance. P comes to < 0.02 on referring to the 't' table. It can occur less than two times in 100, i.e., very rarely by chance. The difference is real in 98% experiments, hence highly significant. In published works, it is written as (t = 2.74, P < 0.02 or significant at 2% level). So vitamins A and D were responsible for the difference in increase of weight in two groups.

2. The erythrocyte sedimentation rate (mm/hour) of 15 male and 10 female TB patients before start of the treatment is given below. (Data sampled from Tuberculosis Chemotherapy Centre, 1956. Bulletin of the World Health Organization 21, 51). Examine the significance of the difference in the means.

The erythrocyte sedimentation rate mm/hr.

Males (X_1)	Females (X_2)
65	63
60	85
115	90
82	100
43	90
103	105
125	98
118	93
83	100
75	125
90	
95	
128	
65	
84	
$n_1 = 15$	$n_2 = 10$
$\overline{X}_1 = 88.73$	$\overline{X}_2 = 94.90$
$SD_1 = 25.3278$	$SD_2 = 15.7653$

Combined variance or SD^2

$$= \frac{\Sigma(X_1 - \overline{X}_1)^2 + \Sigma(X_2 - \overline{X}_2)^2}{n_1 + n_2 - 2}$$

$$= \frac{8980.89 + 2236.90}{23} = \frac{11217.79}{23} = 487.73$$

$$SD = \sqrt{487.73} = 22.08 \quad SE(\overline{X}_1 - \overline{X}_2) = SD\sqrt{\frac{1}{n_1} + \frac{1}{n_2}}$$

$$= 22.08 \times \sqrt{\frac{1}{15} + \frac{1}{10}} = 22.08 \times 0.408 = 9.01$$

$$t_{23} = \frac{(\overline{X}_1 - \overline{X}_2)}{SE(\overline{X}_1 - \overline{X}_2)} = \frac{88.73 - 94.90}{9.01} = 0.68$$

By the other formula,

$$SE = \sqrt{\frac{s_1^2}{n_1} + \frac{s_2^2}{n_2}}$$

$$= \sqrt{42.77 + 24.85} = \sqrt{67.62} = 8.22$$

$$t_{23} = \frac{6.17}{8.22} = 0.75$$

which is almost the same as calculated by other formula.

Referring to table value of 't' at 15 + 10 – 2 = 23 degrees of freedom, the probability of occurrence of calculated value by chance is greater than 0.4 leading to an acceptance of the null hypothesis. Hence, the observed difference between the mean ESRs of the male and female patients is statistically insignificant.

It is advisable that for samples less than 30, former formula using pooled SD be used to find SE.

3. The following table presents the mean cumulative weight loss in grams for 12 patients receiving propranolol and for 11 control patients following sweating during insulin induced hypoglycaemia.

	n	Mean weight loss in gm	SD
Propranolol	12	120	10.0
Control	11	70	8.0

Do the data present sufficient evidence to conclude that the mean cumulative weight loss is different in two groups?

Significance of Difference in Means

To calculate pooled or combined variance, weight loss by each patient is not given, therefore, find the same by the formula.

$$SD = \sqrt{\frac{\Sigma(X - \bar{X})^2}{n - 1}} \text{ or } SD^2 = \frac{\Sigma(X - \bar{X})^2}{n - 1}$$

or $\Sigma(X - \bar{X})^2 = SD^2 \times (n - 1)$

SD and n are given in both the groups. Total sum of squares = $\Sigma(X - \bar{X})^2$ of experiment + $\Sigma(X - \bar{X})^2$ of control
= $(n - 1) \times SD^2$ of experiment + $(n - 1) \times SD^2$ of control
= $11 \times 10^2 + 10 \times 8^2 = 1100 + 640 = 1740$.

Pooled variance or $SD^2 = \frac{1740}{21} = 82.857$

$SD = \sqrt{82.857} = 9.1$

$$SE = SD\sqrt{\frac{1}{n_1} + \frac{1}{n_2}} = 9.1 \times \sqrt{\frac{1}{12} + \frac{1}{11}}$$

$= 9.1 \times \sqrt{0.17424} = 9.1 \times 0.4175 = 3.799$ or 3.8

$$t = \frac{\text{Obs. diff.}}{SE} = \frac{\bar{X}_1 - \bar{X}_2}{SE} = \frac{120 - 70}{3.8} = \frac{50}{3.8} = 13.16$$

Referring to 't' table, corresponding to 21 degrees of freedom, we find that t_{21} at 1% level = 2.831 and t_{21} at 0.1% level = 3.819 showing thereby, that the difference in weight loss with propranolol is highly significant at 0.1% level.

Paired t-test

It is applied to paired data of independent observations from one sample only when each individual gives a pair of observations. Such situations are commonly faced in medical sciences, as in examples below:
1. To study the role of a factor or cause when the observations are made before and after its play, e.g., of exertion on pulse rate; of meals on leucocyte count; effect of a drug on blood pressure; of 'Gugul' an Ayurvedic drug, Bengal gram, Garlic, Onion, etc., on cholesterol levels in the blood.

2. To compare the effect of two drugs, given to same individuals in the sample on two different occasions, e.g., adrenaline and noradrenaline on pulse rate, number of hours for which sleep is induced by two hypnotics and so on.
3. To study the comparative accuracy of two different instruments, e.g., two types of sphygmomanometers.
4. To compare results of two different laboratory techniques, e.g., estimation of haemoglobin by Tallquist and Sahli's methods; microfilaria infection rate by thick smear and concentration techniques; examination of stools for hookworm ova by zinc floatation and concentration methods and so on.
5. To compare observations made at two different sites in the same body, e.g., compare temperature between toes and between fingers or in axilla and mouth, or in mouth and rectum of the same individuals.

Testing by this method eliminates individual sampling variations because the sample is one and the observations on each person in the sample are taken before and after the experiment. In the case of unpaired t-test, there are two samples similar in all respects to eliminate bias, which is often difficult. So pairing is a good idea when the results fall into pairs. A twin is more likely to be like the other one.

One essentially starts with the null hypothesis. It is assumed that there is no real difference between the means of before and after observations.

For testing the significance of difference

1. Find the difference in each set of paired observations before and after $(X_1 - X_2 = x)$.
2. Calculate the mean of the difference ($\bar{x}$).
3. Work out the SD of differences and then the SE of mean from the same, $SD/\sqrt{n}$.
4. Determine 't' value by substituting the above values in the formula,

$$\text{'t'} = \frac{\bar{x} - 0}{s/\sqrt{n}} = \frac{\bar{x}}{s/\sqrt{n}}$$

As per null hypothesis, there should be no real difference in means of two sets of observations, i.e., theoretically it should be 0.
5. Find the degrees of freedom. Being one and the same sample, it should be n − 1.
6. Refer 't' table (Appendix II) and find the probability of the calculated 't' corresponding to n − 1 degrees of freedom.
7. If the probability (P) is more than 0.05, the difference observed has no significance, because such a difference can occur commonly due to chance. Thus, the factor under study may have no influence on the variable. But if P is less than 0.05, the difference observed is significant, because such a difference is less likely to occur due to chance. Influence of the factor to which the sample is exposed may be accepted as an alternative to null hypothesis.

Examples

1. Systolic blood pressure (SBP) of 9 normal individuals, who had been recumbent for 5 minutes was taken. Then 2 ml of 0.5% solution of hypotensive drug was given and blood pressure recorded again. Did the injection of drug lower the blood pressure?

Serial No.	BP before injection X_1	BP after injection X_2	Difference $X_1-X_2 = x$	Squares $= x^2$
1.	122	120	2	4
2.	121	118	3	9
3.	120	115	5	25
4.	115	110	5	25
5.	126	122	4	16
6.	130	130	0	0
7.	120	116	4	16
8.	125	124	1	1
9	128	125	3	9
Total			27	105

Mean difference of $\bar{x} = \dfrac{\Sigma \bar{x}}{n} = 27 \div 9 = 3$

Sum of squares or $\Sigma(x-\bar{x})^2 = (2-3)^2 + (3-3)^2 + (5-3)^2 + (5-3)^2 + (4-3)^2 + (0-3)^2 + (4-3)^2 + (1-3)^2 + (3-3)^2$
$= 1 + 0 + 4 + 4 + 1 + 9 + 1 + 4 + 0 = 24$

$$SD = \sqrt{\frac{\Sigma(x-\bar{x})^2}{n-1}} = \sqrt{\frac{24}{8}} = \sqrt{3} = 1.73$$

By direct formula also SD of differences

$$= \sqrt{\frac{\Sigma x^2 - \frac{(\Sigma x)^2}{n}}{n-1}} = \sqrt{\frac{105 - \frac{27 \times 27}{9}}{9-1}}$$

$$= \sqrt{\frac{24}{8}} = \sqrt{3} = 1.73$$

SE of difference $= \dfrac{SD}{\sqrt{n}} = \dfrac{1.73}{\sqrt{9}} = 0.58$

$t_8 = \dfrac{\bar{x}}{SE} = \dfrac{3}{0.58} = 5.17$

At 8 degrees of freedom, 5% significant limit of 't' is 2.31. The observed 't' value is 5.17 times the standard error hence there is no doubt that the drug injected, produced hypotensive effect, ('t' = 5.17, P < 0.001, highly significant).

2. The weight of 10 tuberculosis patients on admission and at the end of 12 months of treatment with PAS plus isoniazid daily, are given below. The data was sampled from a clinical trial (Tuberculosis Chemotherapy Centre, 1959, Bulletin of the World Health Organization, 21, 51). Examine whether the gain in weight is statistically significant.

Patients' Serial no.	Weight in kg On admission (A)	At 12 months (B)	Weight $X =$ B−A	$X - \bar{X}$ (Mean $\bar{X} = 4$	$(X - \bar{X})^2$
1.	49	52	3	−1	1
2.	41	43	2	−2	4
3.	37	46	9	5	25
4.	41	52	11	7	49
5.	42	46	4	0	0
6.	37	38	1	−3	9
7.	39	42	3	−1	1
8.	38	41	3	−1	1
9.	41	42	1	−3	9
10.	35	38	3	−1	1
Total			40		100

Calculate the difference in weight for each patient. There is a gain of 3, 2, 9, 11, 4, 1, 3, 3, 1 and 3 kg.

The mean gain, X = 4 kg.

$$SD = \sqrt{\frac{\Sigma(X - \overline{X})^2}{n - 1}} = \sqrt{\frac{100}{9}} = \frac{10}{3} = 3.33$$

$$SE = \frac{SD}{\sqrt{n}} = \frac{3.33}{\sqrt{10}} = \frac{3.33}{3.16} = 1.05$$

$$t_9 = \frac{\text{Mean observed difference}}{\text{SE of difference}} = \frac{4}{1.05} = 3.8$$

Referring to table values of 't' at 9 degrees of freedom, the probability of occurrence of this value is less than 0.01. Hence, the mean gain in weight is statistically significant at 1% level.

If number of paired observations is larger than 30, the same method may be used to calculate Z value instead of 't'. Then there is a no need to refer 't' table.

Variance Ratio Test

Comparison of sample variance involves what is called **variance ratio test.** This test involves another distribution called F-distribution. Calculate S_2^2 and S_1^2, i.e., variance of two samples first and then calculate.

$$F = \frac{S_1^2}{S_2^2}$$ (S_1^2 should be greater of the two and be kept as numerator).

The significance of F can be found by referring to F-table. Degrees of freedom will be $n_1 - 1$ and $n_2 - 1$ in the two samples. F-table gives variance ratio values at different levels of significance at df ($n_1 - 1$) given horizontally and ($n_2 - 2$) given vertically.

Example

Apply the F-test of the two samples of 9 and 10 observations of height—

Sample	Sum of squares	df	Variance
1	36	8	36/8
2	42	9	42/9

$$F = \frac{42}{9} + \frac{36}{8} = 1.04$$

This value of F is much lower than the one given in F-table at p, 0.05, therefore it is not at all significant and the samples may very well be regarded as drawn from the same population.

Analysis of Variance (ANOVA) Test

Having understood the variance ratio test we may now discuss the much more important test called the analysis of variance test which is not confined to comparing two sample means, but more than two samples drawn from corresponding normal populations.

Suppose you want to know whether occupation plays any part in the causation of blood pressure. Take BP of randomly selected 10 officers, 10 clerks, 10 laboratory technicians and 10 attendants. Find means and variances of BP of the 4 classes of employees.

If occupation plays no role in the causation of BP, the 4 groups when compared among themselves will not differ significantly. If occupation is playing a significant role, the 4 means will differ significantly.

To test whether the 4 means differ significantly or not, F-test or analysis of variance test has to be applied.

We start with the null hypothesis (H_0) that BP is independent of occupation. To test the hypothesis, proceed as follows:

The first step would be to calculate the sum of squares as given in the example.

Now split this into two components:
i. Sum of squares between the classes
ii. Sum of squares within the classes.

The sum of squares within the classes or groups is found by subtracting sum of the squares between the classes from total sum of squares of the entire sample. Now tabulate the results as below:

Source of variation	df	Sum of squares	Mean sum of squares (MSS) = 3 ÷ 2	F-ratio $\dfrac{MSS_1}{MSS_2}$
1	2	3	4	5
1. Between the classes				
2. Within the classes (Error)				
3. Total				

Compare the calculated F-ratio with that given in the F-table at df between the classes and at df within the classes at 5% level of significance. If calculated value is greater than the table value, H_0 is rejected and H_1 of significant difference between the means is accepted.

If the F-ratio value is smaller than table value, H_0 is accepted and the inference would be that the samples are drawn from the same population.

It may be noted that F-test shows only collective results of all the means. If it is desired to test the difference of any of the two means, then 't'-test may be applied.

Example

Systolic blood pressure values (X) of 4 occupations are given. Determine if there is significant difference in mean blood pressure of 4 groups in order to assess the role of occupation in causation of BP.

	Officers X_1	Clerks X_2	Lab. Technicians X_3	Attendants X_4
	125	120	120	118
	130	122	115	120
	135	115	115	118
	120	110	130	120
	115	125	120	120
	120	122	125	115
	130	120	122	125
	135	120	115	125
	140	126	126	120
	135	120	118	115
Total	1285	1200	1206	1196
$\bar{X}$	128.5	120.0	120.6	119.6

$\Sigma X = 1285 + 1200 + 1206 + 1196 = 4887$

Sum of squares of all the 40 observations

$= 125^2 + 130^2 \ldots, 120^2 + 115^2 = 598751$

Total sum of squares $= \Sigma X^2 - \dfrac{(\Sigma \bar{X})^2}{N}$

$= 598751 - \dfrac{(4887)^2}{40} = 1681.78$

Occupation sum of squares (sum of squares between the classes)

$\dfrac{(\Sigma X_1)^2}{n_1} + \dfrac{(\Sigma X_2)^2}{n_2} + \dfrac{(\Sigma X_3)^2}{n_3} + \dfrac{(\Sigma X_4)^2}{n_4} - \dfrac{(\Sigma X)^2}{n}$

$= \dfrac{(1285)^2}{10} + \dfrac{(1200)^2}{10} + \dfrac{(1206)^2}{10} + \dfrac{(1196)^2}{10} - \dfrac{(4887)^2}{40}$

$= \dfrac{(1285)^2 + (1200)^2 + (1206)^2 + (1196)^2}{10} - \dfrac{(4887)^2}{40} = 538.48$

Error sum of squares (Sum of squares within the classes
= Total sum of squares – occupation sum of squares
$= 1681.78 - 538.48 = 1143.30$.

Analysis of variance (ANOVA) table

Square of variance	df	Sum of square	Mean sum of square	F-ratio
Between the occupation	4−1=3	538.48	179.49	5.65
Error	39−3=36	1143.30	31.76	
Total	40−1=39	1681.78		

Computed 'F' ratio = $\dfrac{179.49}{31.76}$ = 5.65

Table F value for 3 df across (→) and 36 df vertically (↓) at 5% level of significance = 2.86. Since the computed F-ratio is greater than the table F-ratio (critical ratio) the mean BP of the 4 types of employees differ significantly.

On looking at the means for 4 occupations, we find that mean BP of clerks, lab. technicians and attendants are comparable but mean BP is apparently highest in the officers.

Chapter 10

Significance of Difference in Proportions of Large Samples

In the last two chapters, 'Z' and 't' tests of significance of variation in means of large and small samples of *quantitative data were dealt* with. Now *qualitative data* in which the character remains the same while frequency varies will be studied. Number with the same class such as male, rich, survived, vaccinated, cured, died, etc., are counted to form a group.

To study variation in one or more attributes, the data are summarised and expressed mostly as **proportions**. They indicate the relation between the individual events and the events in totality. e.g., if in the total 40 students, 10 are females and 30 are males, the female proportion is 10/10 + 30 = 10/40 or 1:4.

Sometimes the qualitative numbers are expressed as a ratio which indicates relation between two different events such as female to male students in the ratio of 10/30 or 1:3. *Proportions* and *ratios* are comparative measures. If time element enters in, they are expressed as rates such as morbidity, mortality, birth rates, etc., over a period, out of a population of 100, 1,000, 10,000, etc.

If a sample is divided into only two classes such as successes and failures; vaccinated and unvaccinated;

polymorph and non-polymorph (other WBCs); died and survived; etc, it is said to have a *binomial classification* (binomial implies division into two classes). If the sample is divided into more than two classes such as blood groups— A, B, O and AB, or WBCs—polymorphs, lymphocytes, eosinophils, etc., it is said to have a *multinomial* or *polynomial classification*.

The proportion (p) of individuals, having a specific attribute or character in a binomial distribution is expressed as 'P'.

$$p = \frac{\text{Number of individuals having a specific character}}{\text{Total number in the sample}}$$

If p is the probability of occurrence of a positive attribute such as attacked, vaccinated, successes, etc., and q is the probability of non-occurrence of the same or occurrence of alternate attribute such as not attacked, unvaccinated, failures, etc, then q is equal to $1 - p$ in a proportion. If p and q are expressed in percentages then $q = 100 - p$.

Percentage is 100 times the proportion out of one, so to express proportion (p) in percentage, it may be multiplied by 100 as done in expressing probabilities, e.g., 0.05 as 5%.

Examples

1. There were 80 attacks of cholera in a population of 200. Express the data in ratio, proportion and percentage for attacked and escaped.
   i. Ratio of attacked to escaped = a:b or $\frac{a}{b}$ = 80 : 120

   or $\frac{80}{120}$ or 2:3

   (Conventionally a ratio is expressed as

   $1 : \frac{b}{a} = 1 : \frac{3}{2} = 1 : 1.5$)

   ii. Proportion of attacked (p) = $\frac{80}{80 + 120} = \frac{80}{200} = 0.4$

   Proportion of escaped (q) = $\frac{200 - 80}{80 + 120} = \frac{120}{200} = 0.6$

   or $q = 1 - p$ or $1 - 0.4 = 0.6$

   iii. Usually the proportions are expressed in percentages so multiply the proportion by 100.

Percentage attacked, $p = \dfrac{80}{200} \times 100 = 40\%$

Percentage escaped, $q = \dfrac{200-80}{200} \times 100 = 60\%$

or $q = 100 - p$ or $100 - 40 = 60\%$

2. In a class of 88 students there are 66 boys and 22 girls. Express in ratio proportion and percentages.

i. Ratio of boys to girls $= \dfrac{66}{22} = 3:1$

ii. Boys' proportion (p) to total students

$$= \dfrac{66}{66+22} = \dfrac{66}{88} = \dfrac{3}{4} = 0.75$$

Girls' proportion $(q) = \dfrac{22}{66+22} = \dfrac{22}{88} = \dfrac{1}{4} = 0.25$

iii. Boys' percentage $(p) = \dfrac{66}{88} \times 100 = 75\%$

Girls' percentage $(q) = \dfrac{22}{88} \times 100 = 25\%$

or $100 - 75 = 25\%$

STANDARD ERROR OF PROPORTION (SEP)

Standard error of proportion may be defined as a unit that measures variation which occurs by chance in the proportions of a character from sample to sample or from sample to population or vice versa in a qualitative data. It is of particular importance in medical statistics and is frequently used to find the efficacy of a drug, line of treatment, operation, vaccine, etc., or the part played by other casual factors e.g., in the past, 35% cases died in cholera epidemics but 31% died in a new series in the year 1976. As per past experience (population percentage) it should have been 35%. Whether this fall was due to better treatment, change in virulence, less susceptible population or it was only by chance, has to be determined by calculation of standard error of proportion.

The probability or proportional chances of positive or negative occurrence of an attribute or a character in a population or universe, follows Binomial Frequency Distribution. It arises from random sampling of a binomial population. If repeated samples are drawn from such a population, sample proportions (p) follow a normal distribution around the population proportion (P) like normal distribution of means. This simple proportion is used as a point estimator of the population proportion 'P' like normal distribution of mean.

In a binomial distribution of universe, i.e., when only two groups of characteristics are there, if large number of samples are drawn, then the groups in each sample will have a proportion very close to that of the population. Proportions of samples (p) have a central tendency, i.e., a large number of them concentrate around the population (P). The sample proportions (p) are symmetrically distributed around the population proportion (P).

For example, typhoid case fatality rate in a population without treatment was found to be 25%. The mortality percentages (p) of all samples drawn from the same population will have a tendency to concentrate around (P) = 25 and they will be symmetrically or normally distributed.

Normal distribution of sample values around the proportion of universe (P) may be expressed arithmetically in terms of a unit of variation in sample proportions called **standard error of proportion** (SEP) with **binomial confidence limits** as below.

1. 68% of the sample proportions will lie within the limits of population proportion (P) $\pm$ 1 standard error of proportion.
2. 95% values will lie within the range of P $\pm$ 1.96 SE (95% confidence limits). Samples with larger or smaller proportions will be rare, only 5%. Such values are considered statistically significant at 5% level.
3. P $\pm$ 2.58 SE (99% confidence limits) will cover 99% of samples. Thus proportions of samples, falling outside this range, larger or smaller, will be very rare and may be due to play of some influence. It occurs by chance only once in 100 estimates.

Normal distribution of sample proportion (p) around the population proportion (P) may also be expressed graphically by drawing a histogram and curve called *normal curve*, as was done in the case of means and individual observations. It will have the same characteristics described in Chapter 5.

Normal distribution and normal curve are made use of in measuring chance variation of sample proportions from that of universe or variation of one sample value from that of another in terms of standard error of proportion by applying Z-test.

$$Z = \frac{\text{Observed difference}}{\text{SEp}} = \frac{p - P}{\text{SEp}} \text{ (from universe to sample)}$$

or $Z = \dfrac{p_1 - p_2}{\text{SE}(p_1 - p_2)}$ (from sample to sample)

This test is applicable to large samples only and significance of variation is determined at 95% and 99% confidence limits corresponding to Z values of ± 1.96 and ± 2.58, respectively.

In case of small samples the same is applied with a correction factor as per formula

$$Z = \frac{|p_1 - p_2| - \frac{1}{2}\left(\frac{1}{n_1} + \frac{1}{n_2}\right)}{\sqrt{PQ\left(\frac{1}{n_1} + \frac{1}{n_2}\right)}}$$

where $|p_1 - p_2|$ modulus implies that the difference should be taken as positive.

Calculation of SEP

In practice many samples are not taken to find population proportion of a characteristic. To know within what limits it lies from the sample proportion, find the standard error of proportion by the following formula:

$$\text{Standard error of proportion (SEP)} = \sqrt{\frac{p \times q}{n}}$$

where, p is the percentage of positive character, q that of alternate character, and n denotes the number in the sample.

If p is 25% and q is 75% such as deaths and survivals in 100 typhoid cases.

$$SEP = \sqrt{\frac{25 \times 75}{100}} = \frac{25\sqrt{3}}{10} = 4.3$$

Population percentage P will lie—within the range of $25 \pm 1.96 \times 4.3$ in 95% cases, and within the range of $25 \pm 2.58 \times 4.3$ in 99% cases in the above example.

Application and Uses of SEP

1. *To find confidence limits of population proportion (P) when the sample proportion (p) is known* The population proportion will lie in the range of sample proportion $\pm$ 1.96 SE at 95% confidence limit.

2. *To determine if a sample is drawn from the known population or not when the population proportion P is known* If the sample proportion is larger than proportion of known population $\pm$ 1.96 SE, more than 95% chances are that the sample is not drawn from this population. Probability of difference occurring by chance can be found by applying Z-test as done in the case of means.

$$Z = \frac{p - P}{SEp}$$

3. *To find the standard error of difference between two proportions* to know if the observed difference between the proportions of two samples is statistically significant or it is apparent and insignificant due to chance.

4. *To find the size of sample* If the population proportion is known as explained in Chapter 6.

NB—As in Chapter 9 for easy calculations, 2 SEs are taken instead of 1.96 SE for 95% confidence limits and 3 SEs instead of 2.58 for 99% confidence limits in the examples that follow. That makes the text strict, easy to remember and apply.

Examples

1. Find the limits within which you would expect the population proportion to lie if you had examined records of all the 50 children of a school and found 23 had tonsillectomy done.

$p = 23$ out of 50 or $23/50 \times 100 = 46\%$
$q = 27$ out of 50 or $27/50 \times 100 = 54\%$

$$SEP = \sqrt{\frac{p \times q}{n}} = \sqrt{\frac{46 \times 54}{50}} = \sqrt{49.68} = 7.05$$

The 95% confidence limits or population proportion of tonsillectomy done will be $46 \pm 2 \times 7$, i.e., 32% and 60%.

2. Find the population proportion if in a sample of 50 children 25 give history of whooping cough.

$$p = \frac{25}{50} \times 100 = 50 \text{ and } q = 50$$

$$SEP = \sqrt{\frac{50 \times 50}{50}} = 7.08$$

The 95% confidence limits of population proportion will be $50 \pm 2 \times 7$, i.e., 36 and 64%.

3. Polymorph count was 350 out of 500 WBCs. At 95% confidence level, within what limits the population proportion will lie?

$$p = \frac{350}{500} \times 100 = 70 \quad q = 100 - 70 = 30$$

$$SEP = \sqrt{\frac{70 \times 30}{500}} = \sqrt{\frac{21}{5}} = \sqrt{4.2} = 2.05$$

The 95% confidence limits for the population proportion will be $70 \pm 2 \times 2 = 66\%$ and 74%.

4. The proportion of blood group B among Indians is 30%. In a batch of 100 individuals if it is observed as 25%, what is your conclusion about the group?

$$SEP = \sqrt{\frac{25 \times 75}{100}} = \sqrt{\frac{75}{4}} = \sqrt{18.75} = 4.33$$

$$Z = \frac{p - P}{SEP} = \frac{30 - 25}{4.33} = 1.15$$

The difference is insignificant at 95% confidence limits, because 30% is less than the Indian limit of $p + 2$ SE, i.e., $25 + 2 \times 4.3 = 33.6$

5. Constipation was considered to be a common feature as observed in 60% of typhoid cases. In a study of 500 typhoid cases, 30% had constipation. Can you consider

constipation as a common feature of typhoid on this observation?

$$SEP = \sqrt{\frac{30 \times 70}{500}} = \sqrt{\frac{21}{5}} = \sqrt{4.2} = 2.05\%$$

$$Z = \frac{60 - 30}{2.05} = \frac{30}{2.05} = 14.6$$

Difference is highly significant as the proportion (60%) lies much beyond the 99% confidence limit, i.e., $30 + 3 \times 2.05 = 36.15$. In other words, constipation as a common symptom in 60% is a wrong presumption which might have been made from a biased or a small sample.

6. What should be the size of the sample of assessing prevalence rate of diabetics in an urban population where the prevalence was given as 3% in age group above 15 years. Allowable error, L is 10% of positive character with 5% risk.

Prevalence rate 3%.

$$L = \frac{10}{100} \times 3 = 0.3$$

With 5% risk (95% confidence limits)

$$n = \frac{4pq}{L^2} = \frac{4 \times 3 \times 97}{0.3 \times 0.3} = 12933 \text{ persons}$$

Standard Error of Difference Between Two Proportions (SE $(p_1 - p_2)$)

The differences in the pairs of proportions or percentages of samples drawn from the same population are also normally distributed as was seen in case of difference between two means. In actual practice, we do not know the value of population proportion (P) and we have only two samples. So we have to substitute the value noticed in one sample in place of P and compare it with that of the other. The assumptions are:

1. n_1 and n_2 are large
2. Samples are selected at random.

The significance of difference is found by **normal deviate, Z-test.**

$$Z = \frac{\text{Observed difference}}{\text{Standard error of difference}} = \frac{p_1 - p_2}{SE(p_1 - p_2)}$$

If observed difference between the two proportions is greater than 1.96 times the standard error of difference, it is significant at 95% confidence limits, i.e., in 95% cases it would not happen. The chances of its being normal are only 5%. The difference is normal or real and may be due to the influence of some external factor. If the observed difference is greater than 3 times the SE, it is a real variability in 99.73% cases and biological or due to chance only in 0.27% cases.

It may be noted that if a difference between two proportions falls within the range of p ± 2 SE, it does not conclusively prove that the observed difference is only due to play of chance and not due to play of the particular factor under study. This difference could have been due to that factor but it is more likely to be due to chance (Type I error). Similarly, a value of proportion outside the range of P ± 2 SE could be due to chance also but it is less likely to be so. It is more likely to occur due to the factor under study which differentiates the two groups otherwise equal in all respects (Type II error).

Remember that in the tests of significance, one weighs the probability and does not offer mathematical proofs.

Calculation of SE $(p_1 - p_2)$

As for means, the SE of proportion of each sample may be calculated and standard deviation of differences called **standard error of difference between two proportions** denoted as SE $(p_1 - p_2)$ may be found by one of the two formulae that follow:

i. SE of difference is the square root of the sum of the squares of the SEs of the two proportions.

$$SE(p_1 - p_2) = \sqrt{\left(\sqrt{\frac{p_1 q_1}{n_1}}\right)^2 + \left(\sqrt{\frac{p_2 q_2}{n_2}}\right)^2}$$

$$= \sqrt{\left(\frac{p_1 q_1}{n_1} + \frac{p_2 q_2}{n_2}\right)}$$

ii. $SE(p_1 - p_2) = \sqrt{\left(\dfrac{PQ}{n_1} + \dfrac{PQ}{n_2}\right)} = \sqrt{PQ\left(\dfrac{1}{n_1} + \dfrac{1}{n_2}\right)}$

P and Q are combined percentages of positive and negative characteristics in both the samples. So use of combined variance is made. It is simpler for calculation, P is the pooled proportion and Q = 100 − P.

Examples

1. If typhoid mortality in one sample of 100 is 20% and in another sample of 100 it is 30%, find the standard error of difference between two proportions by both the formulae. Is the difference in mortality rates significant?

As per formula (i)

$p_1 = 20$ $q_1 = 80$ $n_1 = 100$
$p_2 = 30$ $q_2 = 70$ $n_2 = 100$

$SE(p_1 - p_2) = \sqrt{\dfrac{20 \times 80}{100} + \dfrac{30 \times 70}{100}} = \sqrt{(16 + 21)}$
$= \sqrt{37} = 6.08$ ...(A)

As per formula (ii)
Total of the two samples = $n_1 + n_2 = 100 + 100 = 200$

$P = \dfrac{p_1 + p_2}{n_1 + n_2} \times 100 = \dfrac{20 + 30}{100 + 100} \times 100 = 25\%$

$Q = \dfrac{q_1 + q_2}{n_1 + n_2} \times 100 = \dfrac{80 + 70}{100 + 100} \times 100 = 75\%$

or Q = 100 − P = 100 − 25 = 75%

$SE(p_1 - p_2) = \sqrt{P \times Q\left(\dfrac{1}{100} + \dfrac{1}{100}\right)} = \sqrt{25 \times 75 \times \dfrac{2}{100}}$
$= \sqrt{\dfrac{75}{2}} = \sqrt{37.5} = 6.12$ ...(B)

A and B are very close to each other.

$Z = \dfrac{\text{Observed difference}}{\text{SE of difference}} = \dfrac{30 - 20}{6.08} = 1.64$

It is less than 1.96, the critical level of significance, hence insignificant at 95% confidence limits.

2. In school A, tonsillectomy had been done in 23 students out of 50 while in the other school B it was done in 77 out of 350. Find if the difference observed in the two schools is by chance or due to some influence.

In school A $n_1 = 50$

p_1 with tonsillectomy = 23 out of 50 = $\frac{23}{50} \times 100 = 46\%$

q_1 without tonsillectomy = 100 − 46 = 54%

In school B $n_2 = 350$

$p_2 = \frac{77}{350} \times 100 = 22\%$ $q_2 = 78\%$

$$SE(p_1 - p_2) = \sqrt{\frac{p_1 \times q_1}{n_1} + \frac{p_2 \times q_2}{n_2}}$$

$$= \sqrt{\frac{46 \times 54}{50} + \frac{22 \times 78}{350}} = 7.39$$

By simpler formula ii

P = 23 + 77, out of 400 = 25%
Q = 27 + 273 out of 400 = 75%

$$SE(p_1 - p_2) = \sqrt{PQ\left(\frac{1}{n_1} + \frac{1}{n_2}\right)} = \sqrt{25 \times 75 \left(\frac{1}{50} + \frac{1}{350}\right)}$$

$$= \sqrt{25 \times 75 \times \frac{8}{350}} = \sqrt{\frac{600}{14}} = \sqrt{42.86} = 6.55$$

$$Z = \frac{\text{Observed difference}}{\text{SE of difference}} = \frac{46 - 22}{6.55} = \frac{24}{6.55} = 3.7$$

Difference is highly significant because at 99% confidence limits it is more than 3 times the SE. It can happen less than once in 100 times by chance. So tonsillectomy is more common in school A. The factor playing part may be investigated.

3. In the school referred to in example 2 above, there was history of whooping cough in 25 students in A and 215 in B. Determine whether the difference is due to chance or real. n_1 in school A = 50, n_2 in school B = 350.

p_1 in school A = $\dfrac{25}{50} \times 100 = 50\%$

p_2 in school B = $\dfrac{215}{350} \times 100 = 61.4\%$

$P = \dfrac{p_1 + p_2}{n_1 + n_2} \times 100 = \dfrac{25 + 215}{50 + 350} \times 100 = 60\%$

$Q = 100 - 60 = 40\%$

$SE(p_1 - p_2) = \sqrt{PQ\left(\dfrac{1}{n_1} + \dfrac{1}{n_2}\right)}$

$= \sqrt{60 \times 40 \left(\dfrac{1}{50} + \dfrac{1}{350}\right)} = \sqrt{54.86} = 7.4$

$Z = \dfrac{p_1 - p_2}{SE(p_1 - p_2)} = \dfrac{50 - 61.4}{7.4} = -1.54$

−1.54 is less than 1.96 SE, hence insignificant at 95% confidence limits.

Chapter 11

The Chi-square Test

Chi-square test, unlike other tests of significance such as 'Z' and 't' tests discussed here-to-fore, is a non-parametric test not based on any assumption or distribution of any variable.

This statistic, though different, also follows a specific distribution known as Chi-square distribution, is very useful in research. It is most commonly used when data are in frequencies such as in the number of responses in two or more categories.

APPLICATION OF CHI-SQUARE

The test involves the calculation of a quantity, called chi-square (χ^2) from the Greek letter 'chi' (χ) and pronounced as 'kye'. It was developed by Karl Pearson and has got the following three common but very important applications in medical statistics as test of:
1. Proportion
2. Association
3. Goodness of fit.

Test of Proportions

As an alternate test to find the significance of difference in two or more than two proportions. In case of large binomial samples of size over 30, the significance could be found by calculating the standard error of difference between two proportions. The method was explained in the last Chapter.

Chi-square test is yet another very useful test which can be applied to find significance in the same type of data with two more advantages.

i. To compare the values of two binomial samples even if they are small, less than 30 such as incidence of diabetes in 20 non-obese with that in 20 obese persons. The test can still be applied provided correction factor, Yates correction, is applied and the **expected** value is not less than 5 in any cell.
ii. To compare the frequencies of two multinomial samples such as number of diabetics and non-diabetics in groups weighing 40 to 50 kg, 50 to 60 kg, 60 to 70 kg and more than 70 kg.

Test of Association

The test of association between two events in binomial or multinomial samples is the most important application of the test in statistical methods. It measures the probability of association between two discrete attributes. Two events can often be studied for their association such as smoking and cancer, treatment and outcome of a disease, vaccination and immunity, nutrition and intelligence, social class and filariasis, cholesterol and coronary disease, weight and diabetes, blood pressure and heart disease, alcohol and gastric ulcer and so on. There are two possibilities, either they influence or affect eath other or they do not. In other words, they are either *independent* of each other or they are *dependent* on each other, i.e., associated. Assumption of independence of each other or no association between two events is made, unless proved otherwise by χ^2 test. Thus the test measures the probability (p) or relative frequency of association due to chance and also if the two events are associated or dependent on each other.

The χ^2 test has an added advantage. It can be applied to find association or relationship between two discrete attributes when there are more than two classes or groups as happens in multinomial samples, e.g., to test the association between number of cigarettes, equal to 10, more than 10, 11 to 20, 21 to 30 and more than 30 smoked per day and the incidence of lung cancer; between incidence of filariasis and social classes—very rich, middle and poor; state of nutrition < 60%, 61 to 80% and 81 to 100% and intelligence quotient, or between parity of the mother 1st, 2nd, 3rd, 4th, 5th and above and the weight of the newborn and so on.

Table 11.1: Outcome of treatment with drug and placebo

Groups	Died	Outcome of result Survived	Total
A. Control, on placebo	10	25	35
B. Experiment, on drug	5	60	65
Total	15	85	100

Table 11.1 is prepared by enumeration of *qualitative* data. *Actual frequencies* as they are, and not the percentage of the characteristics, are entered. Since the table presents joint occurrence of two sets of events, the treatment and outcome of disease, it is called *contingency table* (Latin, *con*, together, *tangere*, to touch).

Since one wants to know the association between two sets of events, the table is called *association table*. When there are only two samples, each divided into two classes, it is also called *fourfold, fourcell* or 2×2 *contingency table* (Table 11.1).

Chi-square (χ^2) test is also applied to find association between two events occurring together when there are more than two classes or more than two samples. Then the contingency or association table would be larger than fourfold or fourcell (Table 11.2).

Table 11.2: Social class and microfilaria positivity

Social class	Number positive	Microfilaria positivity Number negative	Total	Percentage positive
I	4	76	80	5
II	20	180	200	10
III	60	440	500	12
IV	144	576	750	20
Total	228	1272	1500	15.2

Test of Goodness of Fit

Chi-square (χ^2) test is also applied as a test of "goodness of fit", to determine if actual numbers are similar to the expected or theoretical numbers—goodness of fit to a theory. K is the number of classes for χ^2 in goodness of fit test. We can find whether the observed frequency distribution fits in a *hypothetical or theoretical or assumed distribution of qualitative data. Whether or not the observed frequencies of a character differ from the hypothetical or expected ones by chance or due to some factor playing part, can be tested, (such as birth of a male or female child after medicine for wanted sex, male or female—(common fallacy so far).

Assumption of no difference between the observed and hypothetical distribution is made. It is assumed to be 50% in either case. Only insignificant difference may be there. This application is illustrated in the examples No. 7–10.

To test fitness of an observed frequency distribution of qualitative data to a theoretical distribution (normal, binomial or Poisson), again χ^2 test is applied. The test determines whether an observed frequency distribution differs from the theoretical distribution by chance or if the sample is drawn from a different population. It is well illustrated in the example No. 11.

The highest values of χ^2 obtainable by chance are worked out and given in Fisher's table (Appendix III) at different degrees of freedom under different probabilities (p) such as 0.05, 0.01, 0.001, etc. If calculated χ^2 value of the sample is found to be higher than the expected value in the table, at the critical level of significance, i.e., probability of 0.05, the hypothesis of no difference between two proportions or the hypothesis of independence of two characters is rejected. If calculated value is lower, the hypothesis of no difference is not rejected, thereby, concluding that the difference is due to chance, or the two characters are not associated. The exact probability for larger or smaller calculated value of χ^2 is found from the table. It may be 0.1, 0.01, 0.001 or may be somewhere in between any of these two.

---

* Population distribution

172 Methods in Biostatistics

The level of significance of the χ^2 value may be stated in percentages as 5%, 1% and so on, instead of as probability of occurrence by chance out of unity given in the table (P = 0.05 or 0.01 and so on).

NB—While comparing two lines of treatment, an experienced physician may be satisfied with lower standard of significance such as 10% or even 20% though customary level of significance is 5% (thereby committing Type I error). He admits that he could be wrong in 5, 10 or 20 estimates out of 100. While making his choice of level, he has to consider the time, money and material available in clinical practice.

CALCULATION OF χ^2 VALUE

To rule out chance variation in any observed value, apply χ^2 test for which three essential requirements are:
1. A random sample
2. Qualitative data
3. Lowest expected frequency not less than 5. If Yates correction is applied test of proportion can be applied even if expected value is below 30.

If satisfied proceed as follows:
1. Make the contingency tables as above, Tables 11.1 and 11.2. Note the frequencies observed (0) in each class of one event, row-wise, i.e., horizontally and then the numbers in each group of the other event, columnwise, i.e., vertically.
2. Determine the expected number (E) in each group of the sample or the cell of table on the assumption of null hypothesis, i.e., no difference or variation in the proportions of the group from that of the universe.

Total number in the sample is taken as population, such as 1500 slides examined for microfilaria in Table 11.2 out of which 228 were positive. Proportion of total positive to total in all groups or samples is $\frac{228}{1500}$. From this, determine the positive number expected (E_1) out of 80 in the social class I. It would be

$$E_1 = \frac{228}{1500} \times 80 = \frac{182.4}{15} = 12.16$$

Calculate the expected number (E_2) in negative class
E_2 = Total – E_1 = 80 – 12.16 = 67.84; or by the same method.

$$E_2 = \frac{\text{Total negative}}{\text{Total observed}} \times \text{Total in the social class I}$$

$$= \frac{1272}{1500} \times 80 = 67.84$$

Similarly, E values of the other classes II, III and IV are calculated.

3. Find the difference between the observed and the expected frequencies in each cell (O – E).
4. Calculate the χ^2 values by the formula, $\chi^2 = \frac{(O-E)^2}{E}$ for each cell.
5. Sum up the χ^2 values of all the cells to get the total chi-square value,

$$\chi^2_{df} = \Sigma \frac{(O-E)^2}{E}$$

χ^2_{df} indicates the total χ^2 value at particular degrees of freedom indicated at the foot of χ^2 such as χ^2_2.

6. Calculate the degrees of freedom which are related, not to the total number of observations but to the number of categories in both the events. An observer expects no difference between observed and expected frequencies in either direction, vertical or horizontal. So in case of contingency table the formula adopted is

Degrees of freedom (d.f.) = (c – 1) (r – 1)

Where c is the number of vertical columns or classes and r is the number of horizontal rows or groups or classes. In Table 11.2, there are 2 columns of positive and negative classes and 4 rows (I to IV) of social classes or groups.

Hence, df = (2 – 1) (4 – 1) = 3.

Any one expected value is required to be calculated in a 2 × 2 table. Expected values of other 2 cells can be calculated by subtracting it from the column or row total as the case may be.

Former is known as *independent* and the latter as the *dependent* values. Number of independent values will be *one* in 2 × 2 table, *two* in 2 × 3 table, *three* in 2 × 4 table and so on. These independent values, in fact are called the *degrees of freedom*, indicated at the foot of χ^2 such as χ_1^2, χ_2^2, χ_3^2 and so on.

χ^2 is mostly a two-tailed test when you want to find association or test goodness of fit to a theory.

Refer to Fisher's χ^2 table (Appendix IV). Compare the calculated χ^2 value with the highest obtainable by chance at the desired degrees of freedom given in the table under different probabilities such as 0.05, 0.02, 0.01, etc. If calculated value of χ_{df}^2 is higher or lower than the χ_{df}^2 value given in the table, it is significant or insignificant at that particular level of significance to which the reference is made for comparison. Exact level can be determined by comparison with the next higher or lower P value in the table.

Restrictions in Application of χ^2 Test

1. The χ^2 test applied in a fourfold table, will not give a reliable result with one degree of freedom if the expected value in any cell is less than 5. In such cases, to apply χ^2 test, Yates correction is necessary, i.e., reduction of $|O - E|$ by half. In other words, subtract 0.5 from the absolute difference between the observed and expected numbers and square it to divide by expected number. In this case, the formula would be:

$$\chi_1^2 = \Sigma \frac{(|O - E| - \frac{1}{2})^2}{E}$$ (Vertical lines mean modulus, i.e.

it is difference between the O and E values).

2. Even after Yates correction, the test may be misleading if any expected frequency is much below 5. In such cases, some other appropriate test may be applied.
3. In tables larger than 2 × 2, Yates correction cannot be applied. If any cell frequency is less than 5, this small

frequency may be pooled or combined with that in the next group or class in the table.
4. Interpret χ^2 test with caution if sample total or total of values in all the cells, is less than 50.
5. The χ^2 test tells the presence or absence of an association between two events or characters but does not measure the strength of association.
6. The statistical finding of relationship, does not indicate the cause and effect. The χ^2 values do not tell why tonsillectomy and whooping cough, or smoking and cancer are associated or what is the effect of social class on prevalence of filariasis. It only tells the probability (p) of occurrence of association by chance. The conclusions as to these depend on the experimental designs and techniques.
7. Choose the simplest explanation that is in accord with the facts. Hence, if sampling variation can account for certain data, proceed no further.

There are alternative formulae for speedier calculation of χ^2 value, with or without Yates correction in fourfold tables. The same are explained in Table 11.3 below which gives the results of treatment with chloromycetin and furadantin in urinary infection. It has to be found if the results are independent of the drug used.

Table 11.3: Comparative efficacy of the drugs

Drug	Result		Total
	Cured	Not cured	
Chloromycetin	a	b	a + b
Furadantin	c	d	c + d
Total	a + c	b + d	a + b + c + d = N

i. Formula without Yates correction:

$$\chi^2 = \frac{(ad - bc)^2 (a + b + c + d)}{(a + b)(c + d)(a + c)(b + d)}$$

$$= \frac{(ad - bc)^2 \times N}{(a + b)(c + d)(a + c)(b + d)}$$

ii. Formula with Yates' correction:

$$\chi^2 = \frac{(|ad-bc| - N/2)^2 \times N}{(a+b)(c+d)(a+c)(b+d)}$$

|ad – bc| = Modulus (ad – bc) which means the difference ad – bc is to be taken as positive and reduced by (a + b + c + d)/2 before squaring.

In these formulae, expected frequency in each cell is not required to be calculated. Only cross products of observed frequencies in 4 cells have to be calculated, total of each column and row are already there in the table. So they are simpler but the drawback is that they are applicable only to four-fold table while χ^2 test is more often applied to larger tables, when there are many sub-classes and groups in the two events whose association or dependence has to be found.

Examples to demonstrate application of χ^2 test.

1. Apply χ^2 test to find the efficacy of drug from the data given in the Table 11.1 reproduced in Table 11.4. The table is reproduced with expected (E) values noted below the observed (O).

Table 11.4: Outcome of treatment with drug and placebo

Group	Result		Total
	Died	Survived	
A. Control, on placebo	10	25	35
	5.25	29.75	
B. Experiment, on drug	5	60	65
	9.75	55.25	
Total	15	85	100

The expected values (E) are found as explained under step 2 for the calculation of χ^2 value. In simple non-mathematical words:

$$E = \frac{\text{Column or vertical total} \times \text{Row or horizontal total}}{\text{Sample total}}$$

Now find the χ^2 value contributed by each cell by the formula, $\chi^2 = \frac{(O - E)^2}{E}$ and sum up the χ^2 values of all the cells. Total χ^2 value has to be taken into account to draw the conclusion about the efficacy of the drug.

i. Expected number (E) of the 'died' in control group in the Table 11.4 above is

$$\frac{15}{100} \times 35 = 5.25$$

χ^2 value of this cell

$$= \frac{(O - E)^2}{E} = \frac{(10 - 5.25)^2}{5.25} = \frac{22.5625}{5.25} = 4.2978$$

ii. Expected number (E) of the survived in control group

$$= \frac{85}{100} \times 35 = 29.75$$

It may also be found by subtracting the expected number of the died from the total, i.e.,

$$35 - 5.25 = 29.75$$

χ^2 value of this cell

$$= \frac{(25 - 29.75)^2}{29.75} = \frac{22.5625}{29.75} = 0.7584$$

iii. E for the 'died' in experiment group

$$= \frac{15}{100} \times 65 = 9.75$$

χ^2 value of this cell

$$= \frac{(5 - 9.75)^2}{9.75} = \frac{22.5625}{9.75} = 2.3140$$

iv. E for the 'survived' in experiment group = Total – E for died = 65 – 9.75 = 55.25.

χ^2 value of this cell

$$= \frac{(60 - 55.25)^2}{55.25} = \frac{(4.75)^2}{55.25} = \frac{22.5625}{55.25} = 0.4083$$

Total χ^2 value = 4.2976 + 0.7584 + 2.3140 + 0.4083
= 7.7783

On referring to χ^2 table, as 1 degree of freedom, the value of χ^2 under probability 0.05 is 3.841 and under 0.02 is 5.412. Calculated values of χ^2 with and without Yates correction come to 7.778 and 6.227, respectively. Both are higher than 5.41. The difference in mortality is significant at 2% level, hence the drug is efficacious.

2. Attack rates among the vaccinated and unvaccinated against measles are given in the Table 11.5 below. Prove the protective value of vaccination by χ^2 test.

Table 11.5

Group		Result		Total
		Attacked	Not-attacked	
Vaccinated	(O)	10	90	100
	(E)	18	82	
Unvaccinated	(O)	26	74	100
	(E)	18	82	
Total		36	164	200

Expected values may be calculated first and noted in the table.

Total χ^2 value in the 4 cells = $E \dfrac{(O-E)^2}{E}$

$$= \frac{(10-18)^2}{18} + \frac{(90-82)^2}{82} + \frac{(26-18)^2}{18} + \frac{(74-82)^2}{82}$$

$$= \frac{64}{18} + \frac{64}{82} + \frac{64}{18} + \frac{64}{82} = 8.670$$

Though it is a 4-fold table, Yates correction need not be applied as the frequencies are quite large.

Calculated value of χ^2, i.e., 8.670 is greater than the table value 6.64, corresponding to P = 0.01, it is highly significant at 1% level.

3. Determine if there is any association between whooping cough and tonsillectomy when in a random sample of 100 children of a school, 25 had history of tonsillectomy and 60 of whooping cough and 10 had both while 25 had none.

Make the contingency table of the proportions as below (Table 11.6).

Table 11.6

Group	Whooping cough	No whooping cough	Total
Tonsillectomy	O = 10 E = 15 O − E = − 5	O = 15 E = 10 O − E = + 5	25
No tonsillectomy	O = 50 E = 45 O − E = 5	O = 25 E = 30 O − E = − 5	75
Total	60	40	100

$$\chi_1^2 = E\frac{(O-E)^2}{E} = \frac{-5^2}{15} + \frac{5^2}{10} + \frac{5^2}{45} + \frac{-5^2}{30}$$

$$= \frac{150 + 225 + 50 + 75}{90} = \frac{500}{90} = 5.55$$

df = (2 − 1)(2 − 1) = 1

Calculated value, χ_1^2, i.e., 5.55 is greater than the table value of χ_1^2, i.e., 5.41. It corresponds to probability of 0.02 in χ^2, table, hence it is significant at 2% level (P < 0.02). Children with tonsillectomy are likely to get whooping cough in 15 out of 25 and without tonsillectomy 45 out of 75 against 60% and 40%, respectively, as observed. Probability of the difference occurring by chance is less than 2 out of 100 cases, i.e., probability of not getting this difference in nature is 98%. So tonsillectomy and whooping cough are associated or interdependent.

Calculation by alternate formula

$$\chi_1^2 = \frac{(ad - bc)^2 (a + b + c + d)}{(a + b)(c + d)(a + c)(b + d)}$$

$$= \frac{(10 \times 25 - 15 \times 50)^2 \times (100)}{25 \times 75 \times 60 \times 40} = 5.55$$

Thus, the results are same by both the methods.

4. Compare the incidence of accidents in power and hand driven machines. The table of frequencies is given below (Table 11.7).

Table 11.7

Machinery	Accidents	No accidents	Total
Power	8	112	120
Hand	15	165	180
Total	23	277	300

Total χ_1^2 value of 4 cells comes to 0.280

df in this 4-fold table is 1.

At one degree of freedom, χ_1^2 value corresponding to probability 0.05 is 3.841. Calculated value 0.280 is lower, hence not significant at 5% level. Thus, there is no significant difference in the incidence of accidents in two types of machineries.

5. Weights of babies born in the year 1969, in Irwin Hospital, Jamnagar, are tabulated, paritywise (Table 11.8). Is there any association between, parity and weight of babies?

Table 11.8

	No. of babies weighing (kg)					
Mean Weight Parity	1.80 to 2.24	2.25 to 2.69	2.70 to 3.14	3.15 to 3.59	3.60 to 4.04	Total No. of babies
1st	15	25	28	14	2	84
2nd	0	13	20	8	0	41
3rd	0	8	6	9	2	25
4th	0	5	12	10	2	29
5th	0	6	14	11	2	33
6th	0	3	10	7	3	23
7th	2	2	4	14	1	23
8th	0 ⎫	0 ⎫	10 ⎫	5 ⎫	0 ⎫	15 ⎫
9th	0 ⎬ 0	1 ⎬ 2	6 ⎬ 19	4 ⎬ 15	4 ⎬ 7	15 ⎬ 43
10th	0 ⎭	1 ⎭	3 ⎭	6 ⎭	3 ⎭	13 ⎭
Total	17	64	113	88	19	301

Apparently the mean weight rises with the parity indicating association. The number of babies in 8th, 9th and

10th parity being too small may be combined as in Table above. The number of rows is reduced from 10 to 8.

The total χ^2 value comes to 83.043.

Degrees of freedom = $(c - 1) \times (r - 1)$
$= (5 - 1) \times (8 - 1) = 28.$

At 28 degrees of freedom, χ^2_{28} value 83.943, being much higher than the table value of 56.89, is highly significant (P < 0.001). Null hypothesis of no association is rejected and increase in baby weight with parity of mother is proved statistically.

6. Baby weights at birth are given in the following Table 11.9 as per the age of mother. Find if there is any association between the age of mother and the weight of baby.

Table 11.9

Age of mothers in years	No. of babies weighing (kg)					Total No. of babies
	1.80 to 2.24	2.25 to 2.69	2.70 to 3.14	3.15 to 3.59	3.60 to 4.04	
15–20	5	22	29	14	0	70
–25	2	18	35	23	4	82
–30	3	12	27	26	6	74
–35	1	8	20	19	8	56
–40	1	4	9	6	2	22
–45	0	0	0	1	0	1
Total	12	64	120	89	20	305

The total χ^2_{20} value is worked out to be 25.859.

Degrees of freedom = $(5 - 1)(6 - 1) = 20.$

At 20 degrees of freedom, P is > 0.20. It can happen more than 20 times in 100, hence insignificant. So there is no association between the weight of baby and the age of mother.

Examples 7 to 11 that follow, illustrate the application of χ^2 as a test for goodness of fit, in which the degrees of freedom are found by the formula:

df = k − 1 where k indicates the number of classes in the sample on which observations are made.

7. Ratio of male to female births in universe is expected to be 1 : 1. In one village it was found that male children born were 52 and females 48. Is this difference due to chance?

	Male	Female
Observed frequencies	52	48
Expected frequencies	50	50

$$\chi_1^2 = \frac{(52-50)^2}{50} + \frac{(48-50)^2}{50} = \frac{8}{50} = 0.16$$

Significance There are two classes, hence, degrees of freedom = k − 1 = 1 χ_1^2, at 5% level of significance = 3.841. Calculated value of χ_1^2, i.e., 0.16 is much lower, hence insignificant. Observed difference in births of two sexes is due to chance.

8. In a sample of 100 persons, blood group proportions as observed and expected, are given below. Find if the observed distribution fits to the hypothetical (expected) distribution.

	A	B	AB	O
(O)	23	35	5	37
(E)	42	9	3	46

$$\chi_3^2 = \frac{(23-42)^2}{42} + \frac{(35-9)^2}{9} + \frac{(5-3)^2}{3} + \frac{(37-46)^2}{46}$$

$$= \frac{(-19)^2}{42} + \frac{(26)^2}{9} + \frac{(2)^2}{3} + \frac{(-9)^2}{46} = 86.8$$

Significance At k − 1, i.e., 4 − 1 = 3 degrees of freedom, table value of $\chi_3^2 = 7.82$ at 5% level of significance. At the calculated value 86.8, P is far less than 0.001. Hence, the value is highly significant. The observed distribution does not fit to the hypothetical distribution.

9. Nephropathy was observed in 100 cases of each class of diabetics, divided into 4 classes as per severity of disease.

Class	I	II	III	IV
Observed frequencies	8	15	14	7

Is this inequality in different groups due to severity?

According to null hypothesis, severity of disease and incidence of nephropathy are independent so the expected frequency in each class should be the same, i.e.,

$$\frac{8 + 15 + 14 + 7}{4} = 11$$

Class	I	II	III	IV
Observed frequencies	8	15	14	7
Expected frequencies	11	11	11	11

$$\chi_3^2 = \frac{(8-11)^2}{11} + \frac{(15-11)^2}{11} + \frac{(14-11)^2}{11} + \frac{(7-11)^2}{11} = 4.546$$

Significance Degrees of freedom = k − 1 = 4 − 1 = 3. At 3 degrees of freedom table value corresponding to 5% level = 7.82.

The calculated value 4.546 is less, hence, it is not significant. Thus the severity of disease in diabetics does not affect the incidence of nephropathy.

10. Three groups with 20 patients in each were administered analgesics A, B and C. Relief was noted in 20, 10 and 6 cases, respectively. Is this difference due to the drug or by chance?

Group	A	B	C
Observed frequencies	20	10	6
Expected frequencies	12	12	12

Assuming that the type of analgesic makes no difference.

$$\chi_2^2 = \frac{(20-12)^2}{12} + \frac{(10-12)^2}{12} + \frac{(6-12)^2}{12}$$

$$= \frac{(8)^2 + (-2)^2 + (-6)^2}{12} = 8.6$$

Significance Degrees of freedom $k - 1 = 3 - 1 = 2$. At 2 degrees of freedom, at 5% level the value of $\chi_2^2 = 5.99$. The calculated value of χ_2^2 is 8.6, hence significant. The inequality is due to quality of analgesic and not chance.

11. Fit the following observed frequency distribution of students as per height into the Normal Distribution.

Table 11.10

Height in inches	Observed frequency	Expected frequency	Adjusted frequency	O − E	$\chi_2^2 = \frac{(O-E)^2}{E}$
60–63	5	4.2	4	+ 1	0.25000
−66	18	20.6	21	− 3	0.42857
−69	42	39.0	39	+ 3	0.23077
−72	27	28.2	28	− 1	0.03571
72–75	8	7.9	8	+ 0	−
Total	100		100		0.94505

First calculate mean height, $\overline{X}$ and SD, they come to 67.95 and 2.92, respectively.

Next step is to apply the χ^2 test for goodness of fit to work out the normal distribution of 100 students when $\overline{X}$ is 67.95 and SD is 2.92. The same has been done in the Table.

The method adopted is to find the number of individuals who do not exceed the lower and higher values of a group as per unit normal distribution (UND). The difference of the two will give the frequency of that group. As an example, details for the height group 67–72 are given.

For height 69, $Z = \dfrac{69 - 67.95}{2.92} = 0.35$

On referring to the table of UND corresponding to Z value, 0.35, $P = 0.61791 + (0.65542 - 0.61791) + \dfrac{5}{10} = 0.63666$

For height 72, $Z = \dfrac{72 - 67.95}{2.92} = 1.4$

Corressponding to Z value 1.4, the P = 0.91924.

Difference of two values of P, 0.91924 − 0.63666 = 0.28258, is the frequency out of 1. Frequency out of the total 100 = 100 × 0.28258 = 28.258 (adjusted to round figure of 28). In the same way frequencies of other groups are calculated.

χ^2 value corresponding to each group is found from the difference between the observed (O) and expected (E) frequencies. The total χ^2 value comes to 0.94505.

To fit the observed distribution into a normal distribution, 3 degrees of freedom are lost to compensate for 3 restrictions—total frequency, i.e., 100, mean, i.e., 67.95 and SD, i.e., 292. Hence, df = 5 − 3 = 2.

Referring to the χ^2 table, the value at 5% level at 2 degrees of freedom, i.e., χ^2 is 5.99 which far exceeds the observed χ_2^2 value, 0.94505. Thus, the observed distribution fits to the normal distribution. The variation observed is only by chance.

Chapter 12

Correlation and Regression

MEASURES OF RELATIONSHIP BETWEEN CONTINUOUS VARIABLES

So far the quantitative data have been described in terms of averages, dispersion and shape of frequency distribution or curve.

Sometimes two continuous characters are measured in the same person, such as weight and cholesterol, weight and height, etc. At other times, the same character is measured in *two related groups* such as tallness in parents and tallness in children; study of intelligent quotient (IQ) in brothers and in corresponding sisters (siblings) and so on. The relationship or association between two quantitatively measured or continuous variables is called **correlation**. The extent or degree of relationship between two sets of figures is measured in terms of another parameter called **correlation coefficient**. It is denoted by letter 'r'.

When two variable characters in the same series or individuals are measurable in quantitative units such as height and weight; temperature and pulse rate; age and vital capacity; circulating proteins in grams and surface area in square metres; systolic and diastolic blood pressure in mm of Hg, it is often necessary and possible to know, not only whether there is any association or relationship between them or not but also the *degree* or *extent of such relationship*. In case of tachycardia and excitement, the latter cannot be quantitatively described, hence only relationship can be found but not its extent. In the following hypothetical data for the study of correlation between two variables, temperature and pulse rate in 5 persons are given as:

S. No.	Temperature (°F)	Pulse rate
1.	98	72
2.	99	72 + 8 = 80
3.	100	80 + 8 = 88
4.	101	88 + 8 = 96
5.	102	96 + 8 = 104

Careful examination reveals that pulse rises by 8 beats with one degree rise of temperature but such perfect correlation is rarely possible, rather never seen in two biological characters.

Correlation determines the relationship between two variables, but it does not prove that one particular variable alone causes the change in the other. The cause of change in the same or opposite direction may be due to other reacting factor or factors.

The extent of correlation varies between minus one and plus one, i.e., $-1 \leq r \leq 1$.

Correlation between overcrowding and pulmonary tuberculosis is established but it does not mean that high rate of tuberculosis is due to overcrowding alone. Other factors like malnutrition and poor standard of healthy living may be playing a greater role while overcrowding may be a contributory or indirect cause. The χ^2 and correlation coefficient tests help us to disentangle a chain of factors likely to be involved in the causation.

TYPES OF CORRELATION

There are 5 types of correlation depending on its extent and direction. Each type is described mathematically and graphically below.

Perfect Positive Correlation

In this, the two variables denoted by letter X and Y are directly proportional and fully correlated with each other. The correlation coefficient (r) = +1, i.e., both variables rise or fall in the same proportion. Examples of perfect or total correlation are not found in nature but some approaching to that extent are there such as height and weight, age and height and age and weight up to a certain age. X varies directly and proportionately to Y, (X α Y).

The graph forms a straight line rising from the lower ends of both X and Y axes. When scatter diagram is drawn all the points fall on this straight line (Fig. 12.1A).

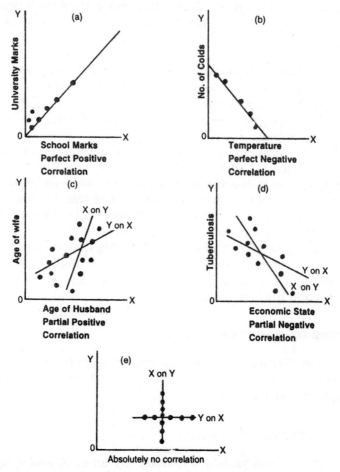

Figs 12.1 A–E: The diagrams are taken from hypothetical numbers to show different types of correlation

Perfect Negative Correlation

Here X and Y values are inversely proportional to each other, i.e., when one rises, the other falls in the same proportion, i.e., the correlation coefficient (r) = –1. Such examples are also not seen in nature but some approaching to that extent are, mean weekly temperature and number of colds in winter; pressure and volume of gas at a particular temperature, etc.

$$X \text{ varies as } \frac{1}{Y} \ (X \alpha \frac{1}{Y})$$

In this case also there will be no scatter, the graph will contain all the observations on a straight line starting from either of the extreme ends because one variable rises and the other falls in a fixed proportion, or r = –1 (Fig. 12.1B).

Moderately Positive Correlation

In this case, the non-zero values of coefficient (r) lie between 0 and +1, i.e., $0 < r < 1$, such as infant mortality rate and overcrowding; filaria incidence and period of stay; temperature and pulse rate; age of husband and age of wife; plasma volume in ml and total circulating albumin in gm; tallness of plants and the quantity of manure used; etc.

In such moderately positive correlation, the scatter will be there around an imaginary mean line, rising from lower extreme values of both variables (Fig. 12.1C).

Moderately Negative Correlation

In this case, the non-zero values of coefficient (r) lie between –1 and 0, i.e., $-1 < r < 0$ such as age and vital capacity in adults; income and infant mortality rate; rainfall and groundnut oil prices in Saurashtra. In moderately negative correlation, the scatter diagram will be of the same type but the mean imaginary line will rise from the extreme values of one variable (Fig. 12.1D).

Absolutely No Correlation

Here the value of correlation coefficient is zero, indicating that no linear relationship exists between the two variables.

There is no mean or imaginary line indicating the trend of correlation. X is completely independent of Y such as height and pulse rate (Fig. 12.1E). The magnitude of correlation coefficients, positive or negative, is indicated by the closeness of dots to an imaginary line indicating scatter or trend of correlation. When points are so scattered that no imaginary line can be drawn, the correlation will be zero.

The closeness of relationship is not proportional to value of r. In other words, if r = 0.6, it does not imply that relationship is twice as close as when r = 0.3. It only shows closeness is more when r = 0.6. Correlation between two variables whether positive or negative, i.e., on plus or minus side, if large, may be significant, otherwise it might have arisen by chance. The significance of the calculated value of correlation coefficient can be computed by finding **standard error of correlation.** Apply 't' test when sample is small. Probability (p) can also be found directly by reference to correlation coefficient table (Appendix V) using the correct number of degrees of freedom (df = n – 2 where n is the number of paired measurements of two variables).

CALCULATION OF CORRELATION COEFFICIENT FROM UNGROUPED SERIES

When associated variables are normally distributed such as height and weight, the correlation coefficient is called **Pearson's correlation coefficient.** It is denoted by symbol 'r' and the following formulae are applied for its calculation. The same cannot be applied when both variables are not normally distributed as in case of intelligent quotient and income because latter is not normally distributed.

$$r = \frac{\Sigma(X - \overline{X})(Y - \overline{Y})}{\sqrt{\Sigma(X - \overline{X})^2 \Sigma(Y - \overline{Y})^2}} = \frac{\Sigma xy}{\sqrt{\Sigma x^2 \Sigma y^2}} \qquad \dots(1)$$

It is laborious because mean has to be found first. Further deduction can be made if the mean is not to be calculated as:

$$r = \frac{\Sigma XY - \frac{\Sigma X \, \Sigma Y}{n}}{\sqrt{\left(\Sigma X^2 - \frac{(\Sigma X)^2}{n}\right)\left(\Sigma Y^2 - \frac{(\Sigma Y)^2}{n}\right)}} \qquad \dots(2)$$

Correlation and Regression

Table 12.1

Child number	Intelligence test score					Marks in examination						
	X	X^2	$\bar{X}$	$X-\bar{X}$ x	x^2	Y	Y^2	$\bar{Y}$	$Y-\bar{Y}$ y	y^2	*XY	xy
1	6	36		−1	1	4	16		−1.5	2.25	24	+1.5
2	4	16		−3	9	4	16		−1.5	2.25	16	+4.5
3	6	36		−1	1	7	49		+1.5	2.25	42	−1.5
4	8	64	56 ÷ 8 = 7	+1	1	10	100	44 ÷ 8 = 5.5	+4.5	20.25	80	+4.5
5	8	64		+1	1	4	16		−1.5	2.25	32	−1.5
6	10	100		+3	9	7	49		+1.5	2.25	70	+4.5
7	8	64		+1	1	7	49		+1.5	2.25	56	+1.5
8	6	36		−1	1	1	1		−4.5	20.25	6	+4.5
Total	56	416		0	24	44	296		0.0	54.00	326	18.0

Values other than X and Y are derived for further calculations from the same.
*Product of intelligence test score and marks in the examination.

This is a simple and direct method like the one used to find the sum of squares in standard deviation. Mean is not required and X and Y are either small measurements when mean is assumed to be zero, or there are differences between large measurements and any other assumed or working mean.

Example

On entry to a school, a new intelligence test was given to a small group of children. The results obtained in that test and in a subsequent examination are tabulated in Table 12.1.
1. Draw a scatter diagram to display the relationship between these two sets of results.
2. Calculate the coefficient of correlation as a measure of closeness of association between the two variables.
3. Derive the regression equation of subsequent examination performance (Y) on initial test result (X). Draw the regression line on the scatter diagram.
4. Comment on the apparent usefulness of the intelligence test in the prediction of examination prowess.

Note that only a small number of observations are provided to make the calculation easier. In practice, many more observations would be needed in an investigation of this kind.
1. Draw the scatter diagram to show apparent correlation if any as per graph Fig. 12.2.
2. Calculate the correlation coefficient by both the formulae and test the significance of the value obtained.
   As per formula (1):

$$r = \frac{\Sigma xy}{\sqrt{\Sigma x^2 \Sigma y^2}} = \frac{18}{\sqrt{(24 \times 54)}} = \frac{18}{36} = 0.5$$

As per simpler formula (2), with X and Y indicating original readings:

$$r = \frac{\Sigma XY - \frac{\Sigma X \Sigma Y}{n}}{\sqrt{\left(\Sigma X^2 - \frac{(\Sigma X)^2}{n}\right)\left(\Sigma Y^2 - \frac{(\Sigma Y)^2}{n}\right)}}$$

$$= \frac{326 - \frac{56 \times 44}{8}}{\sqrt{\left(416 - \frac{56 \times 56}{8}\right)\left(296 - \frac{44 \times 44}{8}\right)}}$$

$$= \frac{326 - 308}{\sqrt{(416 - 392)(296 - 242)}}$$

$$= \frac{18}{\sqrt{(24 \times 54)}} = \frac{18}{36} = 0.5$$

To interpret the observed correlation coefficient (0.5), apply the test of significance, i.e., find the SE of correlation which is a measure of sampling variation.

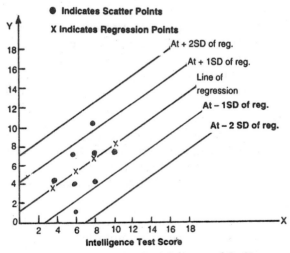

Fig. 12.2: Regression line and standard deviation of the Y measurements

If correlation coefficient (r) in the population is zero, the correlation coefficients of large samples (r) are normally distributed around zero. They will have mean zero and SE,

$$\sqrt{\frac{1}{n-1}}$$

To test the significance, hypothesis of no correlation is followed.

$$\frac{\text{observed r}}{\text{SE}} = \frac{r}{\sqrt{1/(n-1)}} = r\sqrt{(n-1)}$$
$$= 0.50 \times \sqrt{(8-1)} = 0.50 \times \sqrt{7} = 0.50 \times 2.6 = 1.30$$

This value being less than 2 is not significant at 5% level, hence the hypothesis of no correlation is accepted. In other words, intelligence test score and marks in examination have no significant correlation.

This test should be applied to a fairly large sample; here it is done to learn the method, otherwise, significance of the r value of a **small sample** like this should be checked by 't' test.

$$t = \frac{r\sqrt{(n-2)}}{\sqrt{(1-r^2)}} = \frac{0.5\sqrt{6}}{\sqrt{(1-0.5^2)}} = 1.414$$

For $n - 2$, i.e., 6 degrees of freedom, at 5% level, the highest value of 't' obtainable by chance is 2.447.

The estimated value of 't' is 1.414. It is less than 2.447, hence not significant at 5% level ($p > 0.10$). The hypothesis of no correlation is accepted and correlation between intelligence test score and marks in examination is not established. Significance of this sample may also be found by direct reference to correlation coefficient table (Appendix V). For 6 degrees of freedom highest value of r obtainable by chance is 0.707 corresponding to probability of 0.05. Our sample's estimate is 0.50, hence, it is insignificant at 5% level ($p > 0.10$).

Conclusion: The new intelligence test is not useful enough to predict the performance in examination though apparently it may appear to be so.

In a large sample even low degree of correlation (r) may highly be significant while in a small sample high degree of r may be insignificant.

CALCULATION OF CORRELATION COEFFICIENT FROM THE GROUPED SERIES

The data are grouped in correlation table giving two-way frequencies of two variable characters. Such a table is also called *bivariate frequency distribution table*. It classifies quantitative data of two variables, X and Y. Calculation of r from the grouped series is beyond the scope of this book.

When two variables are correlated, but they do not follow normal distribution, another correlation coeficient called **Spearman's rank order correlation coefficient** is used. Its symbol is ρ (Rho).

The formula is

$$\rho = 1 - \frac{6\Sigma d^2}{n(n^2 - 1)}$$ where d is the difference of two corresponding observations in two variables.

Its calculation is beyond the scope of this book.

REGRESSION

In experimental sciences after having understood the correlation between two variables, there are situations when it is necessary to estimate or predict the value of one character (variable say Y) from the knowledge of the other character (variable say X) such as to estimate height when weight is known. This is possible when the two are linearly correlated. The former variable (Y i.e., weight) to be estimated is called dependent variable and the latter (X i.e., height) which is known, is called the independent variable. This is done by finding another constant called **regression coefficient** (b).

Regression means change in the measurements of a variable character, on the positive or negative side, beyond the mean. Regression coefficient is a measure of the change in one dependent (Y) character with one unit change in the independent character (X). It is denoted by letter 'b' which indicates the relative change (Y_c) in one variable (Y) from the mean ($\bar{Y}$) for one unit of move, deviation or change (x) in another variable (X) from the mean ($\bar{X}$) when both are correlated. This helps to calculate or predict any expected value of Y, i.e., Y_c corresponding to X. When corresponding values $Y_{c1}, Y_{c2} \ldots, Y_{cn}$ are plotted on a graph, a straight line called the regression line or the mean correlation line (Y on X) is obtained. The same was referred to as an imaginary line while explaining various types of correlation.

Regression line may be Y line on X, if corresponding values of Y, i.e., Y_c are calculated for X values using the regression coefficient denoted as b_{yx} which is value of y_c, i.e., $Y_c - \bar{Y}$ for one unit of x beyond $\bar{X}$ or vice versa for X on Y.

$$y_c = Y_c - \bar{Y} = b_{yx}(X - \bar{X}) \text{ and } x_c = (X_c - \bar{X}) = b_{xy}(Y - \bar{Y})$$

Regression coefficients of either of the two variables, X and Y, i.e., b_{yx} or b_{xy} for one unit change of the other, can be found by the appropriate formulae.

If height changes by one cm from the mean height $(X - \bar{X} = x$ or $161-160 = 1$ cm) on the baseline, the increment in the weight from the mean weight $(\bar{Y})$ on the vertical line, is calculated by finding regression coefficient b_{yx}. It is the increase in weight in kg, $(Y_c - \bar{Y} = b)$ corresponding to increase in the height (x) by 1 cm from mean $(\bar{X})$.

Correlation gives the degree and direction of relationship between the two variables, whereas the regression analysis enables us to predict the values of one variable on the basis of the other variable. Thereby, the cause and effect relationship between two variables is understood very precisely.

In Figure 12.1 C and D, The two regression lines are shown. One is X on Y and the other is Y on X, indicating conditions of moderately positive and moderately negative correlations, respectively. The two regression lines intersect at the point where perpendiculars drawn from the means of X and Y variables meet.

When there is perfect correlation ($r = +1$ or -1), the two regression lines will coincide or become one straight line (Fig. 12.1 A and B). On the other hand, when correlation is partial, the lines will be separate and diverge forming an acute angle at the meeting point of perpendiculars drawn from the means of two variables. Lesser the correlation, greater will be the divergence of angle. When correlation becomes nil or 0, i.e., the variables are indepedent, the two lines intersect at right angle (Fig. 12.1C).

Steepness of the lines indicates the extent of correlation. Closer the correlation, greater is the steepness of regression lines X on Y and Y on X.

CALCULATION OF REGRESSION COEFFICIENT (B)

Regression coefficient of Y on X is denoted as b_{yx} and that of X on Y as b_{xy}. Regression coefficient of Y for one unit of X, and of X for one unit of Y, are found by either of the following 3 formulae:

a. If correlation coefficient (r) is already calculated the regression coefficient is derived as:

$$b_{yx} = r \times \frac{\text{SD of Y series}}{\text{SD of X series}}$$

$$b_{yx} = r \times \frac{\text{SD of X series}}{\text{SD of Y series}}$$

b. If means are already calculated, the regression coefficients are:

$$b_{xy} = \frac{\Sigma xy}{\Sigma x^2} = \frac{\Sigma(X - \overline{X})(Y - \overline{Y})}{\Sigma(X - \overline{X})^2}$$

$$b_{xy} = \frac{\Sigma xy}{\Sigma y^2} = \frac{\Sigma(X - \overline{X})(Y - \overline{Y})}{\Sigma(Y - \overline{Y})^2}$$

This is an indirect and laborious method.

c. If means are not to be calculated, a simple and direct method is adopted as indicated below:

$$b_{yx} = \frac{\Sigma XY - \frac{\Sigma X \Sigma Y}{n}}{\Sigma X^2 - \frac{(\Sigma X)^2}{n}} \qquad b_{xy} = \frac{\Sigma XY - \frac{\Sigma X \Sigma Y}{n}}{\Sigma Y^2 - \frac{(\Sigma Y)^2}{n}}$$

NB—X and Y are the original measurements or deviations from any assumed mean.

In the above example, the regression coefficient of marks in examination (b_{yx}) for one unit of intelligence is calculated as follows:

$$b_{yx} = \frac{\Sigma xy}{\Sigma x^2} = \frac{18}{24} = 0.75 \quad \text{(indirect method) or}$$

$$b_{yx} = \frac{\Sigma XY - \frac{\Sigma X \Sigma Y}{n}}{\Sigma X^2 - \frac{(\Sigma X)^2}{n}} = \frac{326 - 308}{416 - 392} = \frac{18}{24} = 0.75$$

In this direct method, X and Y being small in size are taken as such.

Regression Line

Regression line for character Y can be drawn on a graph paper by finding expected or calculated values of Y, i.e., Y_c, corresponding to observed values of X from the regression coefficient (b or b_{yx}). Same can be done for X on Y.

Calculated value of Y, i.e., Y_c will be equal to mean of Y, i.e., $\overline{Y}$ plus the regression coefficient (b) multiplied by x, i.e., $(X-\overline{X})$, because if X moves by 1 unit beyond $\overline{X}$ then Y_c moves 1 b beyond $\overline{Y}$ and if X moves by x unit then Y_c beyond Y moves by b × x, i.e., bx units. Thus, calculated value of Y i.e., Y_c corresponding to any value of X is equal to $\overline{Y}$ + bx.

$$Y_c = \overline{Y} + bx = \overline{Y} + b(X - \overline{X}) = \overline{Y} + bX - b\overline{X}$$
$$= \overline{Y} - b\overline{X} + bX$$

$\overline{Y} - b\overline{X}$ being constant may be denoted by a, then $Y_c = a + bX$, 'a' is an intercept of the line, i.e., value of Y when X is zero and b indicates the slope of the line.

Calculate Y_c for the marks corresponding to intelligence test score in the above example as follows:

$\overline{Y} = 5.5$, $\overline{X} = 7$ and b = 0.75

So, constant a = 5.5 – 0.75 × 7 = 5.50 – 5.25 = 0.25.

Now, calculate values of Y, i.e., Y_c corresponding to X values.

Y_c = a + b × X
Y_{c1} = 0.25 + 0.75 × 6 = 4.75
Y_{c2} = 0.25 + 0.75 × 4 = 3.25
Y_{c3} = 0.25 + 0.75 × 6 = 4.75
Y_{c4} = 0.25 + 0.75 × 8 = 6.25

$Y_{c5} = 0.25 + 0.75 \times 8 = 6.25$
$Y_{c6} = 0.25 + 0.75 \times 10 = 7.75$
$Y_{c7} = 0.25 + 0.75 \times 8 = 6.25$
$Y_{c8} = 0.25 + 0.75 \times 6 = 4.75$

Now plot these values (Y_c) of marks corresponding to observed values of intelligence test score, i.e., 6, 4, 6, 8, 8, 10, 8, 6, on the graph paper. A straight line called regression line of Y on X, is formed (Fig. 12.2). It is the mean line of X and Y lines and it will pass through the scatter diagram, about half the points lying on one side and other half on the other. They will be symmetrically distributed like frequencies, if there is a significant correlation. Any expected value of Y can be found on the regression line for given value of X as marks for any intelligence test score. Calculations will not be necessary.

Standard Deviation of the Y Measurements for the Regression Line

Regression line is a graphic test of significance for correlation and can be made use of in finding the extent of correlation between any two observed values from the normal distribution of scatter points around the regression line.

1. As in the distribution of a single set of measurements, the variability of the measurements from a regression line which may be summarised by a variance and standard deviation.
2. Instead of the squared difference between the values and their mean, viz., $\Sigma(Y - \bar{Y})^2$ of a single variable, take the squared difference between the actual and expected or calculated value, viz. $\Sigma(Y - Y_c)^2$.

Find standard deviation of the 'Y' measurements from regression line by any one of the formulae below:

$$\text{SD reg} = \sqrt{\frac{\Sigma(Y - Y_c)^2}{n}} \qquad \ldots(1)$$
$$= \sqrt{\frac{\Sigma(y - y_c)^2}{n}} \qquad \ldots(2)$$

200 Methods in Biostatistics

Table 12.2: Y values

Sl. No.	Y	$\bar{Y}$	Y-$\bar{Y}$ y	Y_c	$Y-Y_c^2$	$(Y-Y_c)^2$	$Y_c-\bar{Y}$ y_c	$y-y_c$	$(y-y_c)^2$
1.	4		−1.5	4.75	−0.75	+0.5625	−0.75	−0.75	0.5625
2.	4		−1.5	3.25	+0.75	+0.5625	−2.25	+0.75	0.5625
3.	7		+1.5	4.75	+2.25	+5.0625	−0.75	+2.25	5.0625
4.	10	44 + 8 = 5.5	+4.5	6.25	+3.75	+14.0625	+0.75	+3.75	14.0625
5.	4		−1.5	6.25	−2.25	+5.0625	+0.75	−2.25	5.0625
6.	7		+1.5	7.75	−0.75	+0.5625	+2.25	−0.75	0.5625
7.	7		+1.5	6.25	+0.75	+0.5625	+0.75	+0.75	0.5625
8.	1		−4.5	4.75	−3.75	+14.0625	−0.75	−3.75	14.0625
Total	44		0			40.5000			40.5000

All values other than Y are derived from Y only.

$$= \sqrt{\frac{\Sigma y^2 - \frac{(\Sigma xy)^2}{\Sigma x^2}}{n}} \qquad \ldots(3)$$

Y_c is taken instead of $\bar{Y}$ to find SD. Firstly because, the calculated values (Y_c) are assumed to be mean values of Y corresponding to specified values of X. Furthermore, the actual values of Y are assumed to have a normal distribution around these mean values of Y_c.

We may find in the example above, the SD of regression for Y_c, i.e., of expected marks in examination from the Table 12.2 of Y values as follows:

Formula (1) and (2) are the same, because $(Y-Y_c) = (y-y_c)$

$Y - Y_c = Y - (\bar{Y} + y_c) = Y - \bar{Y} - y_c = y - y_c$ because $Y - \bar{Y} = y$

Formula (3) is deduced from (2) and is much simpler and if that is applied new Table 12.2 of Y values could be done away with. Insert the values worked out in the formulae, and find SD regression in all, for practice.

$$\text{SD regression} = \sqrt{\frac{\Sigma(Y - Y_c)^2}{n}} = \sqrt{\frac{40.5}{8}} = \sqrt{5.0625} = 2.25 \quad \ldots(1)$$

$$= \sqrt{\frac{\Sigma(y - y_c)^2}{n}} = \sqrt{\frac{40.5}{8}} = 2.25 \qquad \ldots(2)$$

$$= \sqrt{\frac{\Sigma y^2 - \frac{\Sigma(xy)^2}{\Sigma x^2}}{n}} = \sqrt{\frac{54 - \frac{18 \times 18}{24}}{8}}$$

$$= \sqrt{\frac{54 - 13.5}{8}} = \sqrt{\frac{40.5}{8}} = 2.25 \qquad \ldots(3)$$

This being a small sample, divide by $n - 1$, for more accurate result. Then

$$\text{SD regression} = \sqrt{\frac{40.5}{8 - 1}} = \sqrt{5.785} = 2.40$$

Now draw on a graph paper two parallel lines at a distance of one SD regression, i.e., 2.40 from the regression line on either side. Within these lines, 68% of the scatter points will lie. Draw two more parallel lines at a distance of two SD regression. Within these lines, 95% of scatter points of two variables will lie. The same has been done in the Figure 12.2. Marks falling outside ± 2 SD lines, will not be proportional to the corresponding intelligence test score at 95% confidence limits. Either they are too high or too low.

If X line represents the heights and Y line the weights and regression lines are also drawn as in Figure 12.2, the weight of any person can be predicted, corresponding to any desired or specified height and it can also be estimated whether a given weight of a person is proportional to his height or not. If a point for the two variates is plotted, its nearness to regression line will indicate whether it is proportional and its distance will indicate its measure. If the point falls beyond 2 SD regression lines, the body is either bulky or thin and the weight in 95 cases out of 100 is not proportional to height.

In the above exercise, the scatter points appear to be symmetrically distributed though the correlation is not statistically significant. This is because the sample is too small. The data was assumed to be representative of population though it is not so. The sample should be large for valid results and all measurements should be taken under the same conditions and of the same persons.

SUMMARY

1. There are two methods to find whether two variable characters are correlated or not:
    i. *Mathematical* Find the correlation coefficient which is a measure of relationship and the SE to rule out chance.
    ii. *Graphic* Plot the meeting points of the two variable and note the trend in the scatter diagram.
2. The proportionate value of one variable character when it is correlated with another can also be found by either way:
    i. *Mathematically* By finding the regression coefficient first and then the expected or calculated values of one variable, for the observed values of the other.
    ii. *Graphically* By drawing the regression line for values of Y_c corresponding to X values or vice versa. Read

expected value of one, corresponding to observed value of the other directly.
3. Whether a particular value of the variable is proportionate to the other such as height with weight, can also be known:
   i. *Mathematically* By finding, SD of regression coefficient which is a measure of chance variation from regression line or mean line for the two variables X and Y. The proportionate value should lie between the calculated value ± 2 SD regression 95% confidence limits.
   ii. *Graphically* By plotting the value and seeing where it falls, i.e., between the regression line and one SD regression lines or between it and the 2 SD regression lines. The nearer it falls the regression line, the more proportionate it is.

Chapter **13**

Designing and Methodology of An Experiment or A Study

Designing and evolving suitable methodology of an experiment in an institution (hospital or laboratory) or for an inquiry in the field, need very logical and systematic planning. Guidelines on the same are necessary in different fields of medicine for the *postgraduates, research workers* and *field investigators*. They are required to answer questions pertaining to the purpose, scope, objectives, hypothesis, methodology, conclusions and practical applicability of the proposed research. One has to give due consideration to the size and nature of sample, type of matching control, selection of control, etc.

Design consists of a series of *guide posts* to keep one going in right direction and sometimes it may be tentative and not final. During the course of a study, new connecting links in data may also come to light and plan may have to be modified accordingly. Working out of the plan consists of making certain decisions with respect to what, where, when, how much and by what means.

An investigator should take the help of a qualified *biostatistician* right from the beginning and his help should be extended up to the conclusion of study. It has often been observed that biostatistician is contacted only to analyse the data but at this stage his involvement may be too late and cannot compensate for poor planning and collection of improper or inadequate data.

Experts from other fields concerned with the inquiry should also be consulted or involved in planning or conduction of a research study, such as veterinarians, dentists, sanitarians, entomologists, etc.

STEPS IN METHODOLOGY AND DESIGNING

Various steps and methodology usually followed in the design and conduction of an experiment or research project in health and medical practice are described here in brief.

Definition of the Problem

Define the problem you intend to study, such as typhoid mortality after chloramphenicol, iron by injection or by mouth, standard regimen in treatment of tuberculosis, pregnancy prevalence, incidence of viral hepatitis, rising trend in malaria, smoking and lung cancer, cholesterol and coronary heart disease, falling trend in polio cases after pulse polio immunisation use of 75 mg of aspirin in prevention of stroke or reduction in coronary artery block, optimum rural or urban population for a midwife, and so on.

Aims and Objectives

Define the aims and objectives of the study. State whether nature of the problem has to be studied or solution has to be found by different methods, e.g., your objective may be to compare the efficacy of two lines of treatment to adopt better techniques in various methods of heart surgery, modify measures of control if malaria is rising, warn public against smoking if it is proved to be one of the causative factors in lung cancer, recommend active and practical family planning measures if population increase defeats the increase in resources, laparotomy is superior to existing methods of approach in certain abdominal or pelvic operations such as tubectomy, cholicystectomy, etc., and so on.

Review of Literature

Critically review the literature on the problem under study. Find if any such work has been done by others in the past. If so, clarify if you want to confirm the findings, challenge the conclusion, extend the work further or bridge some gaps in the existing knowledge, e.g., density of microfilaria bancrofti is higher at night but you would like to know which part of night, penicillin was found to be very effective as an antibiotic but its toxic or untoward reactions came to light very late, use and abuse of newer antibiotics, different measures of birth control, and so on.

Hypothesis

State your hypothesis After the problem and the purpose are clear and literature on the previous works is reviewed, you have to precisely start with an assumption positive or negative, such as iron by injection is not more effective than by mouth, there is no relationship between hypertension and social status, anopheline mosquitoes are still amenable to DDT, BR and IMR are lower in ICDS projects than in non-ICDS projects or in country as a whole, cholera vaccine is of no use in prevention of cholera, and so on.

Plan of Action

Prepare an overall plan or design of the investigation for studying the problem and meeting the objective The plan should include the following steps:

Definition of Population Under Study

It may be country, state, district, subdistrict, town, village, families or specific groups of population as per age, income, occupation, etc. Define clearly who are to be included and who are to be excluded.

Selection of Sample

It should be unbiased and sufficiently large in size to represent the population under study. Situations in which bias is a problem are:

a. *Retrospective studies* The information recorded cannot be relied upon fully.

b. *Subjective information or data* The patient may tell as it pleases the investigator or the doctor and he may hide or exaggerate certain factors.

c. *Sample not randomly selected* The values of such sample will be biased when applied to population under study.

Size has to sufficiently be large so as to draw valid inferences about the population under study. At the same time, keep in mind, your resources such as money, men and materials as well as the time at your disposal.

Refer to Chapter 6 on "Sampling" for selecting proper technique to choose representative sample for experiment as well as control. To find suitable size of sample, follow the procedure laid down there if the population parameters such as mean, standard deviation or proportion are known from the earlier studies. Otherwise make a pilot study for the problem under investigation in a small population.

Specifying the Nature of Study

Epidemiological studies These are field and not hospital-based. Former studies are more likely to be unbiased while hospital-based studies will be most often biased unless the objective is defined. Infant mortality is higher in rural than in urban areas. Do not infer or presume from the population studied in hospital but go into the respective fields.

a. *Longitudinal studies* They may be prospective or retrospective.

Prospective study means longitudinal follow-up of population over a period of time. It takes time to achieve the objective but will be less biased.

Retrospective (looking back) study is a longitudinal study in profile of a sample or population such as natality performance in women who are 45 years old or over. It involves less expense and gives quick result but is more likely to be biased because it needs rememberance of the past history, which is likely to be forgotten to some extent.

b. *Cohort studies* They are longitudinal studies in which the sample is a cohort.

Cohort is a group of persons exposed to same sort of environment such as newborns, women between 15 and 45 years of age, or workers exposed to radiation or other kinds of hazards in occupation. Cohort study could be prospective such as follow-up of morbidity and mortality in infants from birth to one year of age or it could be retrospective inquiry such as number of persons in the same population who suffered from typhoid in last 5 years.

c. *Interventional studies* In these, there are three phases:
 i. Diagnostic or identification phase,
 ii. Intervention by treatment or service for a specific period,
 iii. Assessment phase for results.

d. *Experimental studies* In these, experiments or trials are made, such as of a drug or some medical services and the results are watched.

e. *Cross-sectional studies* (Non-experimental) It is one time or at a point of time study of all persons in a representative sample of a specific population such as examination of all children in age group 5 to 14 years, detection of cancer cases and study of the factors that lead to cancer, examination of children in age group 0 to 6 years for classifying into nutritional grades, finding prevalence of pregnancy in age group of 20 to 30 years of married women or morbidity due to cancer, paralytic polio, etc. Such studies indicate point prevalence, i.e., number of cases at the time of study. Field surveys in health or disease problems in a community and census are other examples of cross-sectional studies. The study cannot cover the entire population at one point of time. To do that examine part of population today but don't go into the population surveyed already. Next day, study the same items in remaining part, day by day till entire population for study is covered.

Include relevant data about *person, place or time*, e.g. tabulate cancer in smokers and non-smokers as well as in particular age, sex, occupation (persons), prevalence in Bombay and Delhi (place), in the year 1981 and 1991 (period), etc. Such studies are conducted in the field and not in the laboratory or hospital.

f. *Control studies* Most of the experimental studies almost always need a control as a yardstick of evidence. Very rarely, it may not be required as for trial in a fatal disease like rabies. It may be unethical to withhold an established treatment to control cases in which life is at stake or there is a fear of serious after effects. It would not be proper to withhold antibiotics in control cases of typhoid or lobar pneumonia; or a surgical treatment modality like bypass angioplasty in one two or three coronary vessel block, etc.

Sometimes different modalities of cataract operation can be compared where not controls but results will speak.

A control should be identical to experimental group in all respects except for the factor under study. So a matching sample, similar in character and in size, has to be chosen to serve as control. Compare like with the like unless it is deliberate and warranted such as to compare the rate of growth or birth rate in boys and girls.

Controls are ignored in many clinical or field experiments or trials. The conclusions are drawn from a sample of cases with disease, without keeping in view the similar sample of cases with no disease. Have-nots (non-sufferers) form a control for the haves (sufferers), e.g., prevalence of cancer in smokers (sufferers) is compared with prevalence of cancer in non-smokers (control). This is possible only when studies are made in both the groups—experimental and control, similar in all other aspects except the smoking habit.

When vitamins A and D are given to one group of children and not given to another similar group while studying the effect on growth after a specific period, the latter group forms the control. If growth is almost equal in both the cases, only then one can conclude that vitamins have no effect on growth of children.

Different doses of same drug like isoniazid and steptomycin in treatment of tuberculosis may be compared.

Similarly compare three drugs or four drugs treatment of tuberculosis after discovery of *rifampin* for varying lengths of treatment period.

To rule out subjective bias in subjects under study, single or double-blind trial should be made use of.

In a **single-blind trial** one group of the patients is given one drug and another group is given the other drug of the same colour and size or a placebo. So no patient knows what he is being given.

In a **double-blind trial,** not only the patients but also the nurses or medical observers do not know which group of patient or patients are given drug or drugs and which group is on placebo. Placebo and drug are labelled A and B by the principal investigator. Such trials are very useful in comparing effects of two hypnotics or analgesics where subjective information is required.

Selection of patients for treatment and control groups may be done by random numbers. All patients may be collected and distributed two envelopes, red and white at random.

Ruling Out the Observer and Instrument Error

The observers have to be trained. Parallax error in taking readings may creep in due to different positions of the observer. Instruments like weighing machine, sphygmomanometer, height measure, thermometer, etc. should be checked before use. Though very important, this precaution is often ignored.

Recording of Data

For recording data, a standard proforma, schedule, format or a questionnaire has to be prepared and pretested in a few cases. Recorders, interviewers, interrogators and investigators have to be trained in filling these schedules to reduce the personal bias. The following instructions issued to them may be quite useful.
   a. Be friendly and familiar with person or family to be interrogated by proper social approach. Gain confidence

of the interviewee by assuring him that information given will be kept strictly confidential.
b. Questions should be explained clearly and there should be no ambiguity in the reply expected, e.g., age at last birthday or age at the nearest birthday should be specified.
c. Do not ask direct, leading or embarrassing questions such as about fever, disability, pregnancy or sexually transmitted diseases.
d. Too keen or too shy respondents have to be kept in view.
e. Question should be simple to which the reply should be simpler still, like 'yes' or 'no' or 'do not know'.
f. Do not ask too many questions and unnecessary details. Presence of persons other than interviewees may hinder correct response.
g. Questions in the proforma may be open-ended in which the reply is not suggested and empty space is left to record the reply.
h. Proforma may have close-ended questions when respondent has to choose the answers indicated in the format such as for place of birth indicate home or hospital, for birth attendant indicate dai, nurse-midwife, doctor, etc., and so on.
i. Specify to what accuracy an observation is to be recorded, round figures, fractions, etc.

If the proforma is precoded, use standard codes for age groups, sexes, professions, replies to questions, etc., in alphabets A B C .... and or in numbers 1,2,3 ... for data processing and summarising by computers, e.g. Sex: Male—1, Female—2.

Marital Status: Unmarried—1 and so on.

Work Schedule

Prepare work schedule for data collection by estimating work expected per hour, per day, per week or per month, per worker or per team. This should be done by a pilot survey.

Preparation of a **master table:** Each person included in the study is given one line. In this, all subjective and objective information is recorded opposite the identification data like name, age, sex, profession, address, etc.

Presentation of Data

Compile all data and verify their accuracy and adequacy before processing further. Classify and tabulate as per age, sex, class, profession and other desired characteristics. Prepare frequency tables and diagrams as per the type of data.

Unbiased Statistical Analysis

Put the results to unbiased statistical analysis by applying different methods in biostatistics. Statistical machines may be made use of and a statistician, trained in medical biostatistics, should be involved. Discuss the outcome of analysis without preconceived ideas if any.

Conclusions

Draw unbiased conclusions and see if the hypothesis is established as thesis. Re-check the whole plan and its execution before making logical recommendations or preparation of thesis or publication of scientific papers or reports.

Computers

Computers are used in all fields of studies, so as to avoid errors in calculation, for storage data and draw conclusions and in order to save money and time. A separate chapter 18 has been given for their uses in medicine.

SUMMARY

To summarise, after defining the hypothesis and preliminaries for the experiment, there are four stages in study of a problem.

　i. Collection of data from the different sources (involving steps up to the plan of action outlined above).
　ii. Compilation, sorting and presentation of data.
　iii. Analysis of data for measurements of health and disease by the application of statistical methods or techniques.
　iv. Interpretation, drawing conclusions, recommendations and publication of reports.
　v. Use of computers is an asset nowadays in all the above steps.

Now follow a *format* for presentation of any *research work* in the form of a *thesis* for an examination, an *article* in the scientific journal or for a *paper to be read* in a scientific meeting. The format suggested here is only a guideline and it takes into consideration, the various points mentioned in this chapter.

PRESENTATION BRIEFS

1. **Title of the Article** The first step mentioned in the chapter. Next give the name(s) of the author(s) and start the main article with its abstract. The designation, address, etc., are mentioned at the bottom of first page.
2. **Abstract** A brief summary of the work done with the major observations and results is the next item directly below the title and the name(s) of author(s). When printed the abstract is in smaller print or italics or small bold print.
3. **Introduction** Mention here the nature of the problem giving briefly and clearly the reasons for undertaking the particular research. Put forward the hypothesis. Enumerate the aims and, objects of the study.
4. **Review of the Literature** On background of the existing knowledge on the problem chosen for research with relevant references.
5. **Materials and Methods** Steps under plan of action mentioned in the main chapter.
6. **Results** Presentation of data.
7. **Analysis of the Results** Unbiased statistical analysis.
8. **Discussion** In this part, compare the results obtained in your work with those in literature and discuss the reasons for differences and similarities. Also, mention the pitfalls in your study.
9. **Conclusions** A short account of the findings.
10. **Recommendations or Constructive Suggestions**
11. **Acknowledgements**
12. **Bibliography** or **References**
13. **Appendices** (if any).

Further Reading

DB Bisht, Basic Principles of Medical Research.

Chapter 14

Demography and Vital Statistics

Demography is a collective study of mankind. It can be defined as the scientific study of human population, focusing attention on readily observable human phenomena, e.g., changes in population size, its composition and distribution in space. It may be either:

i. **Static demography**—When it means the study of anatomy or structure of communities and their environment in a given population.
ii. **Dynamic demography**—When it deals with physiology or function of communities as regards changing patterns of mortality, fertility and migration. These factors or events change the static structure of the population.

Vital statistics as defined earlier, means data which gives quantitative information on vital events occurring in life, i.e., migration, births, marriages and deaths in a given population. They form essential tools in any demographic study. Vital statistics do not include census and morbidity data but they are explained here for convenience as they are applicable to population changes just like vital events.

COLLECTION OF DEMOGRAPHIC DATA

The different sources for collection of demographic data are:
1. Population census
2. Records of vital statistics
3. Records of health departments
4. Records of health institutions
5. Reports of special surveys

6. Periodic publications by:
   i. World Health Organization (WHO)
   ii. Registrar General, India
   iii. Directorate General Health Services (DGHS), New Delhi
   iv. State Health Directorates
7. Miscellaneous including other health agencies and medical establishments like hospital and nursing homes.

Population Census

It is an important source of health information. The total process of collecting, compiling and publishing demographic, economic and social data at a specified time or times pertaining to all persons in a country or delimited territory, is called a population census (UN Handbook of Population Census Methods).

In good old days, census was carried out to count people for taxation and army recruitment. Now it is carried out to assess the national needs and plan programmes for the people's welfare. Characteristic features of the census are:
1. Full count of population which includes each and every individual and is carried out at regular intervals. In India, the first census was conducted in the year 1872 and the next in 1881. Since then it is being done every ten years and the last one was conducted in 1991.
2. It pertains to a particular territory and the information is collected by making house to house visit on the specified dates in the first quarter of the first year of each decade.

The census in India is conducted under the Indian Census Act, 1948 under which Government is responsible for appointing a Census Commissioner. In practice, the Registrar General of India is assigned an *ex-officio* charge of Census Commissioner. The overall census plan is prepared in conformity with 'Principles and Recommendations for National Population Census, United Nations'.

In the preparatory phase, the Census Commissioner holds elaborate data base conferences and gives a wide publicity to educate people on the importance of census. Preparation of census schedules (proforma) and their pretesting, pilot study and training of enumerators, etc. are carried out. Actual census with a prefixed data (usually March 31) is preceded by house listing operation in which household schedules are canvassed. These operations take approximately one year.

Topics covered are essentially the following:
  i. Total population at the time of census
  ii. Age, sex composition and marital status
  iii. Language spoken, education and economic status
  iv. Fertility, i.e., number of children born alive between two censuses to all women up to the date of census
  v. Citizenship, place of birth, urban and rural populations.

The items in (i) and (ii) are collected invariably in any census while those in (iii), (iv) and (v) are often collected. The changes in the schedule are made from census to census depending on the information needed.

Uses of a Population Census in Health Matters

  i. The census determines population of an area which forms basis for calculation of health indices such as birth, fertility, death, morbidity, marriage rates, etc.
  ii. All health services are planned and community measures are adopted as per nature and size of the population.
  iii. Average rate of growth of population per year is computed from the population enumerated in two census years.
  iv. Knowledge of population distribution helps in planning of the other welfare services such as provision of schools, orphanages, food supplies, etc.

Records of Vital Statistics

Another important source of vital statistics in any country is the Civil Registration System of that country. Civil Registration System in almost all countries of the world has a very long history. Usually it has a legal backing with varying machineries, systems of reporting, proforma for reporting, mode of collecting information, etc.

John Graunt (1620–74), the father of vital statistics analysed London Bills of Mortality, i.e., records of baptism, burials and marriages at the churches, kept by the parish authorities and brought out important facts about births and deaths. William Farr is another great name who started as a compiler in England in 1839, and made notable contributions during 40 years of his service in development of vital statistics as regards notification, registration, analysis and interpretation.

The history of registration in India dates back to the 19th century. Initially only deaths began to be registered with the implicit purpose of assessing the health position. Local Health Officer collected these data and passed them on to the Sanitary Commissioner of the Government of India. The year 1873 can be regarded as the landmark year in the history of Civil Registration in India because it was in that year that Bengal Birth and Death Registration Act was passed. This was later adopted by the neighbouring states of Bihar and Orissa, and was used for formulation of similar Act by other Presidencies and Provinces, in subsequent years. The registration practices and procedures obtaining in different parts of the country varied widely. Nevertheless, with the passage of time, vital statistics became the responsibility of the State Health Directorates. At the grass root level, the local registrars were drawn from either the police department or the revenue department or the education department. Later on, after the advent of Panchayati Raj (1960–62), registration work was also assigned to the secretaries of village panchayats in a few states. In urban areas, the responsibility for registering vital events always rested with Municipal Corporations or Municipalities.

As will be seen from above, there was a great diversity in procedures and practices (including the registration machinery) in different areas of the country. In order to obviate the problems arising from multiplicity of act and rules governing civil registration in different parts of the country,

A Central Legislation on the subject "Registration of Births and Deaths Act" was brought forward in 1969, which makes the registration of births and deaths compulsory within the period specified under rules framed thereunder by the state governments. It came into force from April 1, 1970. It may be clarified that the statutory and executive responsibility for registration activities develops under this Act with the state authorities. The primary motivation for bringing forward the Central Legislation was to introduce standardisation in the concepts and definitions used in collection of information. The Registrar General of India, a statutory authority appointed under this Act, has the responsibility for issuing general directions, and for co-ordinating and unifying the activities of State Chief Registrars. However, the data collection machinery continued to be the same as before, i.e., different state governments have vested the responsibility for registration to different departments, e.g., health, revenue, economics, statistics and with the proviso of a minimum of three-tier heirarchy comprising of a Chief Registrar, District Registrars and Local Registrars. Data on vital events collected from the periphery and the Local Registrars are passed on to District Registrars who, in turn, send them to Chief Registrars. The Chief Registrar of any state or UT has the statutory responsibility for preparing an Annual Report on the functioning of the Civil Registration System, an Annual Statistical Report, and for submission of data to the Registrar General, Statistics India.

The forms and the period for registration of births, deaths, still births or for certification of deaths by the attending doctors have been prescribed under Rules of the State, which are framed to carry out the purpose of the Act. The registration is made compulsory, generally *within 14 days of birth* and *seven days of death*. A default in reporting can attract a fine up to Rs. 50.

Civil Registration System in India, as in most developing countries, has not been functioning satisfactorily and has serious under-recording mainly because of illiteracy in population and lack of much need for having extracts of records. The Civil Registration System in many states does

not net or exceed even 3% of vital events though in some of the progressive states like Kerala, registration efficiency exceeds 80%. It is observed that data obtained from records maintained in routine registers fall short of the actual. In view of the lack of dependable data on Civil Registration System, and non-receipt of returns from the Birth and Death Registrars, the Registrar General, India, has introduced two schemes to monitor the level of registration and its reliability. These are:
1. Sample Registration Scheme (SRS)
2. Model Registration Scheme (MRS renamed as Survey of Causes of Deaths (Rural) in 1982.

Sample Registration System (SRS)

The main objective of SRS, introduced in the year 1964, is to provide reliable estimates of birth and death rates at the state and national levels for rural and urban areas separately. However, SRS also provides information on various other measures of fertility and mortality. The field investigation under SRS consists of continuous enumeration of births and deaths in sampled villages/urban blocks by a resident part-time enumerator preferably a teacher and an independent six monthly retrospective survey by a full-time supervisor. The data thus obtained through these two sources are unmatched and partially matched events are re-verified in the field to get an unduplicated and complete count of all the events. SRS now covers the entire country.

The sample design is random stratified sampling in the rural areas. Stratification is done on the basis of natural division and population size classes. Each natural division within a state has been considered as a stratum and further stratification has been done by grouping the villages into population size classes. In the urban areas, stratification has been done on the basis of population size of cities and towns. The sample unit in rural areas is either a village (if the population is less than 2000) or a segment of a village (if the population is 2000 or above). The sample unit in urban areas is a census enumeration block with an average population size of 750. SRS covers approximately 6 million population in 4149 rural and 1873 urban, i.e., a total of 6022 sample units.

An enumerator is required to send a monthly report to the state head quarters in the first week of the following month. On the basis of the monthly reports received from the sample units, the state headquarters are required to prepare a consolidated montly report and forward the same to the office of the Registrar General, by the end of the following month.

At the national level, the Vital Statistics Division of the office of the Registrar General, India, co-ordinates the implementation work, formulates and prescribes necessary standards, provides necessary instructions and guidance, undertakes tabulation and analysis of data and their dissemination. Latest birth, death and infant mortality rates are given in Tables 15.8 and 15.9).

Ref: Sample Registration System 1984, Vital Statistics Division, Office of the Registrar General, India, Ministry of Home Affairs, New Delhi.

Survey of Causes of Death (Rural)

Collection of mortality statistics forms an integral part of vital statistics system in the country. Mortality influences the growth rate of population and longevity of life. Pattern of deaths by causes reflects the health status of community and, in turn, medical care provided to the community. *Infant mortality rate is the one most important indicator of all-round development of any country.* Thus, information on causes of death is most vital for socio-economic planning.

Due to the continuous paucity of medical personnel and facilities in rural areas where three-fourth of the population of India lives, it is not presently feasible to collect mortality statistics by causes as envisaged in Medical Certification Scheme (MCS).

To bridge this statistical gap on cause of death, the Office of the Registrar General, India initiated Model Registration Scheme (MRS) in 1965. Under this scheme a continuous sample survey of causes of death is being conducted in headquarter villages of selected Primary Health Centres (PHCs) through "lay diagnosis reporting". MRS was renamed in 1982, as Survey of Causes of Death (Rural). The objective of the survey is to build up statistics on causes of death and demographic characteristics (of the deceased) concerning the population by major cause groups by sex and age.

By 1985, the scheme was in progress in headquarter villages of 1080 PHCs, chosen by stratified random sampling technique. The field work for the survey is carried out by trained paramedical personnel of the selected PHCs. Each member of the paramedical staff is allotted a specific segment of HQR village population. Cause of each death occurring in his segment is verified as per criteria outlined in the guidebook, supplied to him. He maintains a prescribed birth and death register.

One member of the staff is identified as "Recorder" who conducts a baseline survey and subsequent half-yearly surveys for obtaining complete count of vital events. He also maintains a national map of the sample village and checks birth and death registers of the field agents (other paramedical workers).

The Medical Officer Incharge of the PHC scrutinises the death register, the symptoms recorded and probable cause of death identified by the field agent. He himself investigates at least 2 deaths or one-tenth of total deaths reported each month by personal visit to the households concerned.

Copies of the entries of the registers of birth and death are forwarded every month from the PHCs to the state headquarters for state-level consolidation of data, monthwise and later transmission to the office of the Registrar General, India.

The implementing agencies for the survey at the state-level are the Directorates of Health Services in some states and Directorate of Economics and Statistics in others. In most cases they are also the Chief Registrar of births and deaths under the Registration of Birth and Death Act, 1969.

The most useful information provided by the survey, relates to the distribution of deaths according to eleven major cause groups of non-medical list. The three digit codes ranging from 001–999 were given to disease categories. This classification was revised by World Health Assembly in 1990–91, which was subsequently amended by DGHS. It is in three volumes and the reader may refer for the same to Central Statistical Organisation, Patel Bhavan, New Delhi.

Definitions of vital events For the sake of comparison at international level, there should be a uniform system of registration and interpretation of vital events. In 1953, the United Nations approved the Principles for a Vital Statistics System representing the consensus of opinion developed over years and published them in 1955, for the guidance of member countries. For details, reader may refer to the Handbook of Vital Statistics Methods, United Nations. Under these principles, there are international definitions of vital events some of which are given below:

1. **Live birth**—Live birth is the complete expulsion or extraction from its mother of a product of conception, irrespective of the duration of pregnancy, which, after such separation, breathes or shows any other evidence of life, such as beating of the heart, pulsation of the umbilical cord or definite movement of voluntary muscles, whether or not the umbilical cord has been cut or the placenta is attached. Each product of such a birth is considered live-born.
2. **Foetal death**—Foetal death is death prior to the complete expulsion or extraction from its mother of a product of conception, irrespective of the duration of pregnancy, the death is indicated by the fact that, after such separation, the foetus does not breathe or show any other evidence of life, such as beating of the heart, pulsation of the umbilical cord, or definite movement of voluntary muscles.

    Foetal deaths are subdivided by period of gestation (measured from the beginning of last menstrual period) as follows:
    i. *Early foetal death*—less than 20 completed weeks of gestation

ii. *Intermediate foetal death*—20 completed weeks of gestation but less than 28
iii. *Late foetal death*—28 weeks of gestation and over.
3. **Still birth**—It is defined as synonymous with late foetal death, i.e., one of twenty-eight completed weeks of gestation or over. It forms an essential part of perinatal mortality.
4. **Immaturity (Prematurity)**—In international classification of diseases an immature infant is a live-born infant with a birth weight of 2500 gm (5½ lbs) or less. In some countries a live-born infant with a period of gestation less than 37 weeks is specified as premature regardless of the weight. Such premature infants may be considered as immature in international classification.
5. **Death**—Death is the permanent disappearance of all evidence of life at any time after livebirth has taken place (postnatal cessation of vital functions without capability of resuscitation).
6. **Infant death**—Deaths occurring under one year of age.
7. **Neonatal deaths**—Deaths occurring under 28 days of age.
8. **Perinatal deaths**—Deaths occurring after 28 weeks of foetal life, in labour and in first week after birth.
9. **Maternal deaths**—Deaths associated with complications of pregnancy, childbirth and puerperium.
(Source WHO TRS No. 25, 1950)

Standard forms for vital events registration The forms vary from country to country but UN Principles for a Vital Statistics System specify certain basic items with priority to be included, to meet the national and international needs. Full name and address of the registered person should be entered for civil needs.
A. **Live Birth Statistics Report** should include:
   a. Data and place of occurrence
   b. Domiciliary (home) or institutional (hospital or maternity home)

c. Sex and legitimacy
   d. Type of birth, i.e., single or plural issues
   e. Attendant at birth such as doctor, midwife, indigenous *traditional birth attendant* (dai) and others
   f. Delivery, normal or abnormal (complication)
   g. Date of registration, to find if it is within specified dates which vary from place to place
   h. Place of usual residence of mother—mothers from villages deliver in the towns for better services.
   i. Date of birth of the mother or her age
   j. Number of children born or birth order, useful for family planning advice.

B. **Death Statistics Report** should include:
   a. Date and place of occurrence
   b. Place of usual residence
   c. Sex and date of birth or age
   d. Cause of death, a very vital item
   e. Certifier, medically qualified or lay person
   f. Date of registration to find time-lag in reporting.

MEDICAL CERTIFICATION OF THE CAUSE OF DEATH

This certification of the cause of death has following objectives:
   a. If underlying cause and chain of events that lead to death are well recorded, effective prevention of precipitating cause from operating is possible.
   b. It has important role in medical research and helps in diagnosis and analysis.

The cause of death is defined as "the morbid condition or disease process, abnormality, injury or poisoning leading directly or indirectly to death. Symptoms or modes of dying such as heart failure, asthenia, etc. are not considered to be the cause of death for statistical purposes".

This does not mean the mode of dying, e.g., heart failure asthenia, etc. It means the disease, injury, or complication which caused death.

Demography and Vital Statistics

International Form of Medical Certificate of Cause of Death (Modified)

Medical practitioner (rubber stamp) or Name of Institution	Allopathic, Ayurvedic, Homeopathic, Unani	Serial number of Institution	Date of Notification
Name of Deceased		Occupation	Date of Death
Address			
Sex	Religion	Age	If under one year Months Days		If within 24 hours Hours Minutes	

Cause of Death		
I		Interval between onset and death, e.g., 5 days
Disease or condition leading to deaths. (This does not mean the mode of dying, e.g., heart failure, asthenia, etc. It means the disease, injury or complication which caused the death.)	(a) Toxaemia due to (or as a consequence of)	
Antecedent cause Morbid conditions, if any, giving rise to the above cause, stating the underlying condition last.	(b) Gangrene of the right leg due to (or as a consequence of)	e.g., 7 days
II	(c) Diabetes mellitus	e.g., 8 years
Other significant conditions contributing to the death, But not related to the disease or condition causing it.	Pulmonary tuberculosis	

Accident, Suicide, Homicide (specify)	How did the injury occur?
Signature of Medical Attendant	Address of the Signatory

Cause of death is an essential part of health statistics. All medical practitioners should use *International Form* (Format attached) after understanding clearly what is meant by a, b and c in Part I and the other conditions in Part II. When there is only one cause, such as an acute infectious disease, it is quite simple, but when there are more than one contributory causes, as in chronic diseases, a uniform procedure has to be adopted.

Part I

Enter at I (a) the disease or condition leading directly to death. This should not be the mode of dying but the disease, injury, or complication which caused death. There must always be an entry at I (a).

If the condition at I (a) was the consequence of another condition, record the next, at I (b). If this, in turn, resulted from an earlier condition, record that condition at I (c).

The fatal sequence will not, of course, always contain three conditions; if the condition at I (a) or I (b) is the underlying cause enter no more in Part I. If the sequence of events comprised more than three stages, extra lines may be added in Part I.

However, if many conditions are involved, write the full sequence, one condition per line starting with the most recent condition at the top and the earliest (the condition that started the sequence of events between normal health and death) last.

The words "due to (or as a consequence of)", which are printed between the lines of Part I, apply not only in sequences with an aetiological or pathological basis but also to sequences where an antecedent condition is believed to have prepared the way for more direct cause by damage to tissues or impairment of function, even after a long interval.

In the case of accident, poisoning or violence, enter a brief description of the external cause on the line immediately below the description of the type of injury or poisoning.

If the direct cause of death arose as a complication of, or from an error or accident in surgery or other procedure or treatment, enter statement of the condition for which the procedure or other treatment was being carried out. (The attending doctor must, of course, comply with any local regulations for the referral of deaths due to violence, etc. to the coroner or other legal authority.)

Normally, the condition or circumstances mentioned on the lowest line used in Part I will be taken as the basis for underlying cause though classification of it may be modified to take account of the complications or other conditions, entered by special provisions of the international classification of deaths.

Part II

Enter in order of sequence, all other diseases or conditions believed to have unfavourably influenced the course of the morbid process and thus contributed to the fatal outcome but they were not related to the disease or condition causing death.

There will be cases where it will be difficult to decide whether a condition relevant to death should be recorded as part of the fatal sequence in Part I or as a contributory condition in Part II. Conditions in Part I should represent a distinct sequence so that each condition may be regarded as being the consequence of the condition entered immediately below it. Where a condition does not seem to fit into such a sequence, consider whether it belongs to Part II.

Interval between onset and death Space is provided, against each condition recorded on the certificate, for the interval between the presumed onset and the date of death. This should be entered where known, at least approximately or "unknown" should be written. This provides a useful check on the sequence of causes as well as useful information about the duration of illness in certain diseases.

The cause of death has to be kept confidential. In the developing countries, the cause mentioned by laymen is usually a symptom such as fever, senility, etc., which has no significance.

The certifiers can also be classified as:
a. Medical categories such as medical attendant, medical examiner and pathologist.
b. Non-medical categories such as midwife, nurse, coroner, sanitary inspector and layman.

As much coverage as possible for certifying the cause of death by doctor or paramedical personnel should be attempted to enhance the value of death records. The death certificates are then scrutinised at the Health Bureau (or an intermediate tabulating agency if any and the causes of death are tabulated according to the underlying cause. In order to ensure comparability with other states and countries the latest Ninth Revision of the International Classification of Diseases, Injuries and Cause of Deaths, a 3 digit classification 000 to 999, prepared by the WHO is used. Students should glance at this publication in their college library. It is in two volumes. Volume I explains the international form of death certificate, and contains rules for classification. Volume II has an alphabetical list of diagnostic terms and contains towards the end an intermediate list of 150 causes of death and an abbreviated list of 50 causes.

The medical practitioner is not concerned with tabulation. This is done by the specially trained clerical staff. But the practitioner must understand how to fill the International Form of Death Certificate. This is in common use today but quite often, it is filled wrongly. A few minutes study would avoid this.

Collection of vital statistical data above the state level

The Bureau of Health Statistics in the State Directorate of Health Services submits the vital data collected, to the Director General of Health Services of India, Delhi, who in turn, furnishes the same to WHO Regional Office for South-East Asia in New Delhi and from there the information goes to WHO Head Office, Geneva and to UN for international study.

Shortcomings in registration of vital statistics

The defects are considerably reduced after the registration responsibility was passed on to the Panchayats and passage of

Birth and Death Registration Act, 1969 under which the recording of vital events has been made compulsory. Still much remains to be done to reduce omissions, avoid duplicate recording, and to record right cause of death, etc. Role of medical personnel is very important in collection of health statistical data and the same is discussed eventwise.

Birth Medical man in general practice or service in the PHC or elsewhere, as well as paramedical person is very often an attendant on an antenatal case, at birth or in postnatal period and when he treats the morbidity in infants. He is conversant with the mode of notification and its importance. He should advise registration of birth in the interest of his patient and health statistics.

Traditional birth attendants (dais) conduct most of the deliveries especially in the rural areas and urban slums. They should be trained to notify each birth, still birth, and paid some incentive for reporting each event.

Services of the medical practitioners should be utilised on part-time payment basis or they should be paid some incentive for each and every event notified. They should be given printed forms with stamped and addressed envelopes for reporting births, deaths and illnesses. They should be supplied with a copy of the International Classification of Diseases, Injuries and Causes of Deaths which they should follow while issuing death certificates. The local health authority should keep uptodate lists of all practitioners in modern and indigenous medicines. Doctors may be approached individually as well as through their associations for help in improving the health statistics data.

Records of Health Departments

Notification of epidemic diseases such as cholera, smallpox and plague is compulsory in rural India by the village officers such as Secretary of village Panchayat to the Taluka office, who conveys the same by telegrams to Collector, District Health Officer and Director of Health Services, for immediate action. In Municipalities and Corporations, notification is compulsory for the doctors attending on a case of a notifiable disease or by the head of the family, or the landlord, to the local health authority and for that there are rules. The data are usually incomplete, and often unreliable for want of correct diagnosis of the disease reported, and under registration. Collection is done at the state level from where the reports are sent to the World Health Organization (WHO). The list of notifiable diseases varies from area to area.

Records of Health Institutions

The routine records include returns from hospitals, health centres, dispensaries and maternity homes. The institutions record all indoor and outdoor cases that come for treatment, their diagnosis and result. Monthly returns are sent to the State Director of Health Services. These data are biased because they do not cover any specified area or population. So they cannot be used as a general measure of morbidity. However, they reflect a gross picture of diseases prevalent in the catchment area of the institutions. Such data are mostly used to assess the hospital needs or efficacy of certain therapeutic measures. A sample form of monthly report to be submitted by a hospital is appended herewith (Table 14.1).

Reports of Special Surveys

Well conducted and comprehensive surveys give useful data about the health status of the community and the progress made to the measures adopted. Epidemiological surveys about malaria incidence, combined with entomological surveys about vector, breeding places, etc., before and after the insecticide such as DDT spray gives useful data on malaria morbidity. Comprehensive surveys are carried out for the assessment of tuberculosis, leprosy and cancer prevalence. Sample surveys are made for finding the correct position when birth and death registration is not reliable. Appraisal of nutritional status is made by physical examination and diet surveys. Specified sample surveys are the only answer to diagnose many of the country's health problems when existing registration is not reliable.

Table 14.1: Format for monthly hospital abstract

Records	Current month	Same month last year
I. Total beds available		
II. Total No. of patients discharged:		
a. Adults and children		
b. Newborn infants		
III. Days of care to patients discharged:		
a. Adults and children		
b. Newborn infants		
IV. Average length of stay based on days of care to patients discharged:		
a. Adults and children		
b. Newborn infants		
V. Total deaths including newborn:		
a. Deaths under 48 hours		
b. Deaths over 48 hours		
c. Gross death rate in percentage		
d. Net death rate in percentage		
VI. Total No. of patients admitted:		
a. Adults and children		
b. Newborn infants		
VII. Daily census of hospital patients:		
a. Maximum on any one day		
b. Minimum on any one day		
c. Total No. of patients cared for in hospital (as per daily census)		
i. Adults and children		
ii. Newborn infants		
VIII. Daily average number of patients:		
a. Adults and children		
b. Newborn infants		
IX. Average percentage of bed occupancy		
a. Adults and children		
b. New born infants		
X. Birth in the hospital:		
a. Male babies		
b. Female babies		
XI. Type of admission during the month (based on the discharged)	This month	Same month last year
Routine		
Emergency		

XII. Geographical distribution of patients admitted:
   Delhi
   Haryana
   and so on

XIII. Operations conducted   Major
   Name of the department   Minor
   Surgery
   Obstetrics and   (Give
      Gynaecology   departmentwise
   Ear, Nose and   figures)
      Throat
   and so on

XIV. Departmentwise distribution of admissions and discharges for the month

Department	Admissions	Discharges	Days of care	Average length of stay	Total deaths	Deaths under 48 hrs.	Deaths over 48 hrs.	Deaths gross %	Net death %
Medicine									
Paediatrics	(Give departmentwise figures)								
and so on									

XV. Patients attendance at various general OPDs/Special Clinics

Name of Discipline	Current	Month	Same month	Last year
	New	Old	New	Old
Medicine,				
Skin and STD				
Paediatrics	(Give disciplinewise figures)			
and so on.				

Others
Casualty, EHS, Nutrition, Family Welfare—Sterilisation, IUD, oral pills, MTP, etc., as per local requirement.

The essential requirements of a good survey are:
  i. A well defined objective
  ii. A well drafted plan of operations and questionnaire
  iii. An epidemiologist or health expert
  iv. Adequate statistical facilities
  v. Sufficient trained and willing staff for the field, with other needed facilities such as good transport, medicines, to get good cooperation, etc.

vi. A fairly large representative or random sample, chosen by sampling techniques (Refer Table 2.14 on tuberculosis).

Periodic Publications

A. *Publications of WHO*

Those helpful in the follow-up studies of health statistics are:
   i. Demographic Year-book of UNO which is a comprehensive collection of demographic statistics
   ii. UN Principles for a Vital Statistics System
   iii. Handbook of Population Census Methods
   iv. The Technical Report Series, containing Reports of Expert Committees on Health Statistics
   v. The Annual Epidemiological and Vital Statistics, in which annual figures from 1939 onwards have been published
   vi. The Monthly Epidemological and Vital Statistics Record
   vii. The Weekly Epidemiological and Vital Statistics Record
   viii. The Annual Report of the Director General which includes an account of statistical activities
   ix. Bulletin of Regional Health Information, SEA No. 1986.

B. *Publications of the Registrar General, India*
   i. Sample Registration System
   ii. Sample Registration Bulletin (Biannual)
   iii. Registrar General's Report on the Working of Registration of Births and Deaths Act, 1969
   iv. Vital Statistics of India
   v. Mortality Statistics of India
   vi. Survey of Causes of Deaths (Rural)
   vii. Socio-economic Statistics India, by Director General, Central Statistical Organisation July 1996.

C. *Publications of the Directorate General of Health Services, New Delhi*

    i. Health Services in India
    ii. Principal Causes of Death in India
    iii. Epidemiological Intelligence System in India
    iv. Health Information of India (CBHI, DGHS)
    v. Family Welfare Programme in India (Year-book).

D. *Publications of State Health Directors*

Their most important publications are the Annual Reports which give lot of vital data concerning their own population.

Miscellaneous Including Other Health Agencies

Data on the causes of death, recorded by insurance companies may be useful but is insufficient. Morbidity and mortality data recorded by some of the industrial managements with, well developed factory health departments are very useful to assess the health of the worker and the occupational hazards.

COMPILATION AND PRESENTATION

The Vital Statistics data after collection are compiled, tabulated and presented systematically at different levels such as District, State, National and International as per their needs.

All the reports are arranged areawise and their adequacy, completeness and accuracy of details are scrutinised in the Bureau of Statistics. The information is transferred to punch cards by electric punchers, then sorted by electrically operated sorters. This minimises the errors and also saves labour. Manual sorting by cards with marginally punched holes, slotting those with positive characteristics, is out of date but still suitable for small data. **Computers** have facilitated tabulation and revolutionised analysis still further (see Chapter 18).

Methods of Tabulation and Presentation of data have been described in Chapter 2. United Nations Organization has suggested that marital status may be classified as single, married, widowed and separated.

Age Classification

The WHO Regulations, Article 6 stipulates that one of the following age groupings shall be used:
1. For general purposes
    i. Monthwise under 1 year, yearwise up to 5th year, five years, groupwise from 5 to 84 years, 85 years and above
    ii. Under 1 year, 1–4 years, 5–14 years, 15–24 years, 25–44 years, 45–64 years, 65 years and over
    iii. Under 1 year, 1–14 years, 15–44 years, 45–64 years, 65 years and over.
2. For specific purposes such as infant mortality rate detailed break-up under one year is recommended as below:
    i. 0 to 7 days (Perinatal Mortality)
    ii. 0 to 28 days (Neonatal Mortality)
    iii. 29 days to 1 year (Post-neonatal Mortality).

This could also be subdivided into mortality from 1 to 5 months and from 6 to 11 months. This may be necessary because the causes of death vary in these subgroups depending on agent, host and environmental factors.

NB—Please refer to Chapter 2, for proper mode of recording of age groups till old age.

Chapter **15**

Measures of Population and Vital Statistics

The objectives of measuring population, vital statistics and morbidity in the field of preventive medicine and public health are:
  i. To *calculate indices* that measure levels of health and morbidity in a population.
  ii. To *compare natality, mortality or morbidity* of different places, times, communities and professions. Comparison may also be useful as per age, sex and different socio-economic conditions.
  iii. To *monitor* the *progress* made in the *health* and *family welfare programmes*.
  iv. To fix up "*priorities*" in adoption of future health measures based on feed-back on the nature of the problem.
  v. To determine the expectation or *longevity of life* at birth or at any particular age, and also to determine the chances of *survival* or the proportion or percentage of survivors at any specific age following an intervention or therapy in a particular disease, e.g., cancer.

MEASURES OF POPULATION

Mid-year Population

The main source of population and its socio-demographical composition is **census,** carried out every 10 years (last in 1991) by Census Commissioner of India, New Delhi. It forms the basis for projecting population data for any year in

Measures of Population and Vital Statistics

between the two census years, e.g., birth rates for the years 1975, 1990, 2000, etc. (Tables 15.1 and 15.2, Fig. 15.1).

Table 15.1: Decennial population in India*

Year	Population (Millions) as per census	Estimated rates in the decenniums 1901-11 onwards		
		Birth rate	Death rate	Growth rate per 1000
1901	238.3	49.2	42.6	6.6
1911	252.0	48.1	47.2	0.9
1921	251.3	46.4	36.3	10.1
1931	278.9	45.2	31.2	14.0
1941	318.6	39.9	27.4	12.5
1951	361.0	41.7	22.8	18.9
1961	439.2	41.2	19.0	22.2
1971	548.1	37.2	15.0	22.2
1981	685.1	33.9	12.5	21.4
1991	855.7	29.5	9.8	19.7
1992	NA	29.2	10.1	19.1
1993	NA	28.7	9.3	19.4
1994	NA	28.6	9.2	19.4

* Source: Registrar General, India, Selected Socio-Economic Statistics, India, 1996.

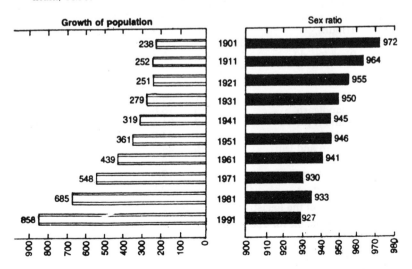

Fig. 15.1: Trends in population of India 1901 to 1991, growth of population and sex ratio

Table 15.2: Statewise population and vital statistics of India 1991

S.No.	State	Population (millions)	Birth rate	Death rate	IMR 1994*	Sex ratio F/1000 males
1.	Andhra Pradesh	66508008	23.7	8.3	63	972
2.	Arunachal Pradesh	864558	27.4	9.3	NA	860
3.	Assam	22414322	30.7	9.1	77	923
4.	Bihar	86374465	32.5	10.4	66	911
5.	Goa	1169793	14.3	6.5	NA	967
6.	Gujarat	41309582	27.1	8.7	64	935
7.	Haryana	16463648	30.5	7.8	67	865
8.	Himachal Pradesh	5170877	26.2	8.6	67	976
9.	J and K	7718700	NA*	NA*	NA	923
10.	Karnataka	44977201	24.9	8.1	65	960
11.	**Kerala**	**29098518**	**17.3**	**6.0**	**16**	**1036**
12.	Madhya Pradesh	66181170	32.8	11.5	98	931
13.	Maharasthra	78937187	24.9	7.4	54	934
14.	Manipur	1837149	21.1	6.6	NA	958
15.	Meghalaya	1774778	29.5	7.1	NA	955
16.	Mizoram	689756	NA*	NA*	NA	921
17.	Nagaland	1209546	19.0	4.2	NA	886
18.	**Orissa**	**31659736**	**28.0**	**11.1**	**103**	**971**
19.	Punjab	20281969	25.0	7.6	53	882
20.	Rajasthan	44005990	33.7	8.9	84	910
21.	Sikkim	406457	24.6	2.9	NA	878
22.	Tamil Nadu	55858946	19.0	7.9	59	974
23.	Tripura	2757205	21.9	5.3	NA	945
24.	Uttar Pradesh	139112287	35.4	11.0	88	979
25.	West Bengal	68077965	25.1	8.3	61	917

Union Territories

1.	A and N Islands	280661	18.0	3.2	NA	818
2.	Chandigarh	642015	18.6	3.6	NA	790
3.	D and N Haveli	138477	34.4	9.4	NA	952
4.	Delhi	9420644	24.8	6.7	70	872
5.	Daman and Diu	101586	24.2	5.9	54	969
6.	Lakshadweep	51707	26.3	7.1	NA	943
7.	Pondichery	807785	17.8	7.5	NA	979
8.	All India	846302688	28.6	9.2	73	927

Source: Office of the Registrar General, India.

* Data not available for 1991.

Measures of Population and Vital Statistics

To calculate indices of health, the total events in a particular year such as births, deaths and morbidity of a defined geographical area form the *numerator* and the population exposed in that year on first July or mid-year population as popularly called, forms the *denominator*. This ratio may be multiplied by 100, 1000, 100,000, etc., to get the figure of convenient magnitude for easy comparison.

There are three methods for calculation of **mid-year** *population:*

A. Natural Increase Method

Natural increase implies the difference between increase in population due to births and immigration and decrease due to deaths and emigration over a period of time say 1 year or 6 years. This can be added to the previous census population to get the mid-year population of a particular year up to which the net increase has been calculated.

This requires reliable recording as done in developed countries like England and Wales but not in developing countries like India. Moreover, by this method, the postcensal or future population cannot be projected.

B. Arithmetical Progression or AP Method

Here it is assumed that there is equal increase each year and average increase per year is calculated in the intercensal period of 10 years such as between 1981 and 1991.

Population of any year or up to a particular month = last census population + period in years and months after the last census month and year × yearly increase.

This method is adopted for its being simple to find intercensal population. It is like simple interest and pays no regard to fluctuations in births, deaths, immigration and emigration. To calculate the health indices for any year, mid-year population is used as denominator, though population for any month in the year can be calculated if need be. The formula applied is:

$$P_t = P_0 + rt$$

Where t is the period in years after the last census, P_t is the population at the required time, i.e., 't' years after the last census, P_0 is the population of last census and r is the annual increase rate.

Example

The population of Delhi as per 1st March, 1981, was 62,20,000 and as per 1st March, 1991, census was 94,21,000. Calculate the mid-year population in 1996 and 2001.

Increase in 10 years = 94,21,000 – 62,20,000 = 32,01,000
Average increase per year = 3,20,100

To determine the mid-year population of 1996 add the increase at the rate of 3,20,100 for 5 years and 4 months from 1st March, 1991 to June 30, 1996.

a. Mid-year population in 1996

$$= 94,21,000 + 3,20,100 \times 5 + 3,20,100 \times \frac{1}{3}$$

($\frac{1}{3}$ means for 4 months from 1st March to 30th June)

$$= 1,11,28,200$$

Similarly, to calculate mid-year population in year 2001, add the increase at the rate of 3,20,100 for 10 years and 4 months from 1st March, 1991 to June 1991.

b. Mid-year population in year 2001

$$= 94,21,000 + 3,20,100 \times 10 + \frac{3,20,100}{3}$$

$$= 94,21,000 + 3,20,1000 + 1,06,700$$

$$= 1,27,28,700$$

C. *Geometrical Progression or GP Method*

It is based on the principle, population begets population. In AP method constant increase throughout was assumed while in this, assumption is made that the percentage or per person rate of growth remains constant. It is calculated like compound interest. If a population grows from one million to three millions in one year, next year it grows from three millions to nine millions and so on, i.e., as per geometric law of growth.

Suppose P_0 is the population of any census year and r is the annual increase per person in intercensal years and P_t (population after t years) is to be found. The formula applied is:

$$P_t = P_0 (1 + r)^t$$

At the end of the year it becomes $P_0 (1 + r)$

At the end of two years = $P_0 (1 + r) (1 + r) = P_0 (1 + r)^2$

At the end of ten years = $P_0 (1 + r)^{10}$

So, at the end of t years = $P_0 (1 + r)^t$

It is easy to calculate with the help of log tables.

Examples

1. Census population of **Delhi** on 1st March, 1981 and 1991 was 62,20,000 and 94,21,000, respectively. Calculate the mid-year population of Delhi in 1996 and 2001.

If 'r' denotes the rate at which the population increases in geometric progression per year, the first step would be to calculate 'r'. Population has gone up in 10 years from 62,20,000 to 94,21,000.

Population on 1st March, 1991 = Population on 1st March, 1981 × $(1 + r)^{10}$

$$94,21,000 = 62,20,000 \times (1 + r)^{10}$$

a. Taking logarithm

$$\log 94,21,000 = \log 62,20,000 + 10 \log (1 + r)$$

$$\log 9,421 + 3 \log 10 = \log 622 + 4 \log 10 + 10 \log (1 + r)$$

$$3.9741 + 3 = 2.7938 + 4 + 10 \log (1 + r)$$

$$\log (1 + r) = 0.01803$$

After March 1991 in 5 years 4 months, the mid-year population in 1996 = $94,21,000 (1 + r)^{5 + \frac{1}{3}}$

Taking log, we get
Log (mid-year population in 1996)

$$= \log 94,21,000 + \left(5 + \frac{1}{3}\right) \log (1 + r)$$

$$= 6.9741 + \frac{16}{3} \times .01803$$

$$= 7.0702$$

Taking antilog, mid-year population becomes

$$= 1,17,60,000$$

b. Similarly, the mid-year estimated population in 2001

$$= 94,21,000 \, (1 + r)^{10 + \frac{1}{3}}$$

Taking logarithms
log (mid-year population in 2001) = log 94,21,000
$+ \left(10 + \frac{1}{3}\right) \log (1 + r)$

$$= 6.9741 + \frac{31}{3} \times .01803$$

$$= 7.1604$$

Taking antilog, we get mid-year population in 2001

$$= 1,44,60,000$$

2a. The population of India as per 1st March, 1981 was 685 million and as per 1st March, 1991 census, was 844 million. Calculate the mid-year population in 1996 and 2001.

Increase in 10 years = 844 − 685 = 159 million

$$\text{Average increase per year} = \frac{159}{10} = 15.9 \text{ million}$$

To determine the mid-year population of 1996 add the increase at the rate of 15.9 million for 5 years from 1st March, 1991 to 1st March, 1996.

$$\text{Mid-year population in 1996} = 844 + 15.9 \times 5 + \frac{15.9}{3}$$

$$= 923.5 \text{ million} + 5.3$$
$$= 928.8 \text{ million}$$

Similarly, to calculae the mid-year population in **2001** add 15.9 per year for 10 years from 1st March, 1991 to 1st March 2001 plus increase of 4 months till mid-year in the end of June.

Mid-year population in 2001

$$= 844 + 15.9 \times 10 + \frac{15.9}{12} \times 4 = 948.3 \text{ million}$$

2b. Census population of India on 1st March, 1981 and 1991 was 685 million and 844 million, respectively. Estimate the mid-year population of India for 1996 and 2001.

If 'r' denotes the rate at which the population increases in geometric progression per year, the first step would be to calculate 'r'. Population has gone up in 10 years from 685 million to 844 million.

Population on 1st March, 1991 = Population on 1st March $1981 \times (1 + r)^{10}$

$$84,40,00,000 = 68,50,00,000 (1 + r)^{10}.$$

Taking logarithm on both sides, we get

$\log 84,40,00,000 = \log 68,50,00,000 + 10 \log (1 + r)$
$= \log (844 \times 10,00,000) = \log (685 \times 10,00,000) + 10 \log (1 + r)$
$= \log 844 + 6 \log 10 = \log 685 + 6 \log 10 + 10 \log (1 + r)$
$= \log 844 = \log 685 + 10 \log (1 + r)$
$2.9263 = 2.8357 + 10 \log (1 + r)$

$$10 \log (1 + r) = 0.0906 \Rightarrow \log (1 + r) = 0.00906$$

After March 1991 in 5 years the mid-year population in $1996 = 84,40,00,000 (1 + r)^5$

Taking log,

log (mid-year population in 1996)

$$= \log 844 + 6 \log 10 + 5 \log (1 + r)$$
$$= 2.9263 + 6 + 5 \times 0.00906$$
$$= 8.9716$$

Taking antilog, the mid-year population of India in 1996 comes to, 936.7 million.

Similarly, the mid-year estimated population of 2001

$$= 84{,}40{,}00{,}000 \, (1 + r)^{10}$$

Taking log
log (mid-year population in 2001)

$$= \log 844 + 6 \log 10 + 10 \log (1 + r)$$
$$= 2.9263 + 6 + 10 \times 0.00906$$
$$= 9.0169$$

Taking antilog, the *mid-year population of India in 2001 comes to 1,04,00,00,000, i.e., 1040 million.*

The reader is notified that the above calculated population is till the March of 1996 and 2001. He should add population increase for 4 months by the same method using $t = \frac{1}{3}$ years and add to the above calculated values to get exact mid-year population till the end of June 1996 and 2001, respectively.

Based on principle of compound interest, geometrical progression or increase is a better method to follow. It gives higher estimate of postcensal and lower estimate of intercensal period than AP method. The same is noticed in examples 1 and 2, where postcensal population of Delhi or of India, respectively in 1996, is found to be higher, and in the intercensal, will be found lower if calculated in 1976.

Growth of Population (Natural Increase)

It is often required to calculate the yearly growth rate of population in demography and family planning. It may be found by subtracting annual death rate from birth rate per thousand. Sometimes it is expressed in percentage.

The population in India has been rising by 2.23% per year in 1981–91 decennium against the annual rise of 2.22% and 1.89% in the previous two decades, i.e., 1971–81 and 1961–71, respectively (Tables 15.1 and 15.3, Fig. 15.2).

Measures of Population and Vital Statistics 245

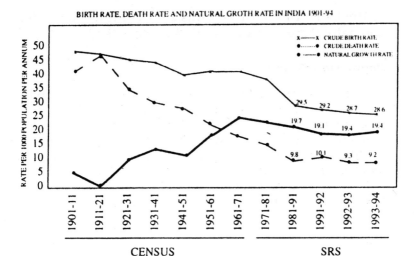

Fig. 15.2: Natural growth rate decennial from census to census and in years 1992–94

Table 15.3: Quinquennial projections of population (000's) by age and sex as on 1st March 1981–2001 India (medium projection)

Year	Persons	Male	Female	Sex ratio males/1000 females
1981	685.18	357.38	333.77	1071
1986	761.07	392.79	368.28	1066
1991	837.28	431.33	405.95	1062

It is seen from the Tables 15.1 and 15.3, and Figure 15.2, that decline in death rates is much smaller as compared with increase in birth rates; nearly 16.06 million people were being added every year in the decennium 1981–91. In 1966, estimated population was 500 million. With birth rate of 40, there were 20 million births and with death rate reduced to 16, there were 8 million deaths. Thus, the increase in population went up by 12 million giving a growth rate of 24 per thousand. As per Sample Registration Scheme in India, in 1993, the crude birth rate was 28.5, crude death rate 9.2 and natural growth rate was 19.3 per thousand.

Population Density

It is defined as number of people living in one square kilometre area. It is a good indicator to assess the slow or rapid increase in congestion due to population growth over different periods of time. With an unplanned increase, the health of the people is affected and, thereby, morbidity and mortality rates go up.

Population density has to be compared among different towns or regions to study the congestion.

Example

Area of India is 31,60,789 sq kilometres and the population in 1981 and 1991 censuses was 683 and 843 millions, respectively. Calculate the rise in population density.

$$\text{Density in 1981} = \frac{68,33,29,097}{31,60,789} = 216/\text{sq km}$$

$$\text{Density in 1991} = \frac{84,39,30,861}{31,60,789} = 267/\text{sq km}$$

Rise in density = 267 − 216 = 51/sq km in the decade.

Density in 1901, was 77 per sq km. With increase in population decadewise, it rose to 82, 81, 90, 103, 117, 142, 177, and 216 in 1981. In the year 1991, it went up to 267 (Source: Provisional Population Tables. Census of India. Paper I, 1991).

Density of population in a town may be calculated as persons per room to determine overcrowding. Density ratio of over 2 persons per room is indicative of overcrowding. *Studies have revealed that general morbidity and infant mortality rates are correlated with overcrowding.*

Population Distribution by Age and Sex in India

Table 15.4 and Figure 15.3 give agewise and sexwise distribution of population in the years 1981 and 1991 in India.

Table 15.4: Percentage distribution of population by sex and age groups

Age group in years	1971			1981			1991		
	Persons	Males	Females	Persons	Males	Females	Persons	Males	Females
0–4	14.5	14.2	14.9	12.6	12.3	12.9	13.1	13.1	13.0
5–9	15.0	14.9	15.1	14.1	14.0	14.1	11.7	11.8	11.6
10–14	12.5	12.8	12.2	12.9	13.1	12.6	11.5	11.6	11.3
15–19	8.7	8.9	8.4	9.6	9.9	9.4	10.8	11.2	10.4
20–24	7.9	7.6	8.1	8.6	8.4	8.8	9.6	9.4	9.7
25–29	7.4	7.2	7.8	7.6	7.5	7.8	8.2	8.1	8.3
30–34	6.6	6.4	6.8	6.4	6.3	6.5	7.0	7.0	7.0
35–39	6.0	6.1	5.9	5.9	5.8	5.9	6.1	6.1	6.2
40–44	5.2	5.3	5.0	5.1	5.3	5.0	5.0	5.0	5.0
45–49	4.2	4.4	3.9	4.4	4.5	4.3	4.5	4.4	4.5
50–54	3.7	3.9	3.6	3.8	4.0	3.6	3.6	3.6	3.6
55–59	2.3	2.4	2.3	2.5	2.5	2.5	3.1	3.1	3.1
60–64	2.6	2.6	2.6	2.7	2.7	2.7	2.2	2.1	2.2
65–69	1.3	1.3	1.3	1.4	1.4	1.5	1.9	1.8	2.0
70+	2.1	2.0	2.1	2.4	2.3	2.4	1.9	1.8	2.1
All ages	100.0	100.0	100.0	100.0	100.0	100.0	100.0	100.0	100.0

Source: Office of the Registrar General, India

248 *Methods in Biostatistics*

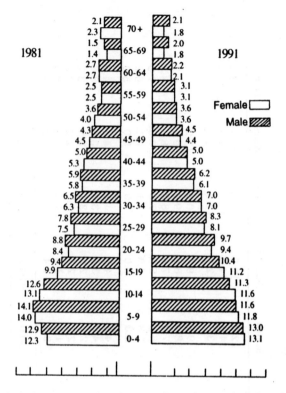

Fig. 15.3: Age-sex distribution of population of 1981 and 1991

In the year 1981, it is evident that the sex-age distribution is constantly tapering towards old age because of high birth rate which gives a broad based conical population pyramid. In developed countries, the pyramid is spindle-shaped, showing larger adult population in the middle and smaller children and elderly population at each end. Due to socio-economic changes, rising age of marriage and popularity of fertility control measures, a similar trend is seen in India in age-sex distribution in the years 1971, 1981 and 1991 as per Table 15.4 (Fig. 15.3). Tendency to decrease in child population below 5 years and increase in older population above 60 years is seen in both the censuses. Tables 15.3 and 15.4, and Figure 15.1 indicate that the proportion of females is lower than males throughout while in developed countries and **Kerala state** in India, the number of females is higher. In 1991, there were 929 females/1000 males (Fig. 15.1).

Higher population of females and low of males are best indicaters of all round progress of any nation.

MEASURES OF VITAL STATISTICS

Absolute number of vital events such as births and deaths are converted into rates and ratios for comparison of vital statistics from place to place or year to year, e.g., infant deaths in U.P. and Kerala.

Rate

Rate is a proportion with a defined denominator termed as population. The numerator is occurrence of an event which is counted in the same population. Multiply by 100, 1000, 10,000, etc. as per magnification desired. When occurrence of an event is rare such as cancer, multiply by 1,00,000. *Rate differs from proportion in the matter of time.* No time factor is involved in the proportion. The rate is a measure of the speed at which new events are occurring in a community. The rates, in general, are of two types, crude and specific.

i. *Crude Rate*

Crude rates are general rates calculated without paying regard to specific sections of the population. They measure the proportion between total events and the total population over a period of time.

Crude rate

$$= \frac{\text{Total number of events that occurred in a given geographical area during a given year}}{\text{Mid–year population of the geographical areas for the same period}} \times 1000$$

ii. *Specific Rate*

This rate is for specific group of the population such as for a particular age, sex, marital status, occupation, etc.

Specific rate

$$= \frac{\text{Number of events which occurred among a specific group of the population of a given geographic area during a given year}}{\text{Mid-year population of the specific group of the population in the same geographic area during the same period}} \times 1000$$

The above rates may have to be calculated for any period of time but most commonly they are expressed on annual basis. Annual rates are found for the calendar year. The adjustment is made for computing annual rates by dividing by an *adjustment factor* such as 5 or 10 to get the annual rates.

Quinquennial rate

$$= \frac{\text{Number of vital events which occurred among the population of a given geographic area during the period of five years}}{\text{Population estimated at the middle of the five years}} \times 1/5 \times 1000$$

Decennial rate

$$= \frac{\text{Number of vital events which occurred among the population of a given geographic area during the period of ten years}}{\text{Population estimated at the middle of the ten years}} \times 1/10 \times 1000$$

Annual rates from the events recorded for the periods shorter than a calendar year are expressed again on annual basis by multiplying with the adjustment factor.

Weekly rate It may be required to compare the rates in different weeks of the year as per season.

Weekly rate

$$= \frac{\text{Number of vital events occurring in one week}}{\text{Total mid-year estimated population}} \times 52 \times 1000$$

Monthly rate The adjustment is made by multiplying that month's rate, with 12 or more correctly, with 365 divided by the number of days in that month.

Quarterly rate The adjustment factor will be 4, i.e., multiply the quarterly rate with 4.

Half yearly rate The adjustment factor will be 2.

NB—In the formulae that follow, the geographical area and the period (mostly a calendar year) in which the events occurred, remain the same both for numerator and denominator, hence, they are not specified again and again for the sake of simplicity.

Measures in the form of rates and ratios may be considered for births, marriages, morbidity and deaths in that order.

Measures of Fertility

i. *Crude Birth Rate*

It is the most commonly used composite index of increase in population by natality or child birth. The formula is:

Annual crude live birth rate

$$= \frac{\text{Number of live births which occurred among the population of a given geographic area during a given year}}{\text{Mid-year population of the same geographic area during the same year}} \times 1000$$

Note—'Live' is distinct from 'total births' which include still births too.

Crude birth rate of India in 1981, was 33.9, in 1985, it was 32.9 (rural 34.5 and urban 28.1) and in 1992, it was 29.2 (rural 30.9 and urban 23). The measure is a crude one because total population taken as denominator is not exposed to births, only married women, 15 to 44 or 49 years of age are exposed. Natality is affected by several factors such as age and sex composition, marriage rate, economic factor, etc., which are not regarded in this rate.

Example

In India, the mid-year population was 605 million and 21 million births took place in 1976. Calculate the birth rate.

$$BR = \frac{2,10,00,000}{60,50,00,000} \times 1,000 = 34.71 \text{ per } 1,000$$

Accordingly calculated BR of India in 1991 was 29.5. Note the fall due to F.P. Programme

Weekly, monthly, quarterly or half-yearly birth rates are calculated after multiplying with the **adjustment factor** explained above, to compare with birth rates in other periods of the same years, or corresponding periods of the previous years.

ii. *Fertility Rate*

This rate is related to women in reproductive period hence it is better to measure the ferdility of population in relation to the number of women aged 15 to 49 instead of the whole population.

a. *General fertility rate*

General fertility rate (GFR)

$$= \frac{\text{Number of live births in one year}}{\text{Number of women aged 15–49 years}} \times 1,000$$

For India, it was 153.1, 120.7 and 145.2 per 1000 women (15–49) in the year 1984, for rural, urban and total population, respectively (SRS). In 1000 population surveyed around Sevagram by the author it was found to be 150 in the year 1976, as against 156.6 in rural India.

b. *General marital fertility rate* In this, the denominator is only the married women in age group 15–49. In India, it was found to be 170.2 in rural and 143.6 in urban population in 1978.

c. *Age specific fertility rate*

Age specific fertility rate

$$= \frac{\text{Number of live births to mothers of a specified age group}}{\text{Mid–year female population of same age group}} \times 1,000$$

By this, the fertility rate of any age group or year of life can be calculated. Its use is made in monitoring family planning measures as given for rural population only in Table 15.5 for years 1981 and 1991. The table shows that fertility rate is highest in age groups 21–24 and 25–29 and then it tapers towards the age group 45–49. It is low in age group 15–19 but still higher than in the age groups after 35–39.

Table 15.5: Age specific fertility rates—India

Age group (years)	1981	1991
15–19	90.4	76.1
20–24	246.9	234.0
25–29	232.1	191.3
30–34	167.7	117.0
35–39	102.5	66.8
40–44	44.0	30.6
45–49	19.6	12.1
Total fertility rate	4.5	3.6

Source: Registrar General India, 1994

Total fertility rate From age specific fertility rates, a further measure called the total fertility rate (TFR) can be calculated. It gives an estimate of the average number of *children born to a woman throughout her reproductive age* as per prevailing specific fertility rates. It was *4.5 per women in 1981* and came down to *3.6 in 1991* (source same as for Table 15.5).

iii. *Reproduction Rate*

Number of births in a population depends upon the number of women in reproductive age. The reproduction depends on the number of female children born. There are two indices of reproduction.

a. *Gross reproduction rate (GRR)* In this, total female children born to women in cohort sample is counted presuming that all of them live up to the entire period of reproductive age.

$$\text{GRR} = \frac{\text{Total number of female children born to women in the cohort sample}}{\text{Total number of women in the same cohort sample}}$$

It was 1.7 and 2.3 in the year 1984, for urban and rural areas, respectively, combined 2.2 per women (source as for Table 15.5). Instead of cohort in which 30 years follow-up is required, it may be calculated from single year age-specific fertility rates considering only female births by the life table method.

b. *Net reproduction rate (NRR)* Women of reproductive age as well as the children born to them are exposed to the mortality experience. In this, find the number of women surviving out of the cohort at the end of the reproductive period and also determine the number of female children left when the cohort women reached the end of the reproductive period as per their mortality experience (life table method). Then, divide the number of girls survived by the number of women survived at the end of their reproductive period. That gives the net reproduction rate (NRR).

$$\text{NRR} = \frac{\text{Number of girls survived after the mortality experience}}{\text{Number of cohort women survived at the end of reproductive period as per their mortality experience}}$$

If the value of this index is one, the population is maintaining itself and if it is more or less than one it means the population is increasing or decreasing. NRR is a demographic indicator. The present NRR was 1.48 in 1981. Target for 2006 AD is NRR = 1.

iv. *Sex Ratio at Birth*

It is expressed as number of females per 1000 males and calculated as:

$$\text{Sex ratio at birth} = \frac{\text{Number of female live births}}{\text{Number of male live births}} \times 1{,}000$$

In 1991, sex ratio was 929 females per 1000 male births.

v. *Illegitimacy Rates*

The two indices are:
Illegitimate fertility rate

$$= \frac{\text{Number of illegitimate live births}}{\text{Unmarried and widowed women population of age group 15–49}} \times 1{,}000$$

$$\text{Illegitimacy ratio} = \frac{\text{Illegitimate live births}}{\text{Legitimate live births}} \times 100$$

Measures of Marital Status

General or crude marriage rate is usually calculated per 1000 population for any year.

Crude marriage rate

$$= \frac{\text{Marriages registered in a calendar year}}{\text{Mid-year population of the same year}} \times 1{,}000$$

Sex ratio among ever married and unmarried rates are based on the population of a particular sex above the marriageable age. Under the existing law in India, the minimum marriageable age for women is 18 years and for men it is 21 years. Age and sex group has to be stated for numerator and denominator.

Female unmarried rate

$$= \frac{\text{Total unmarried females in age group 15–49 years in a population}}{\text{Total females in the same age group}} \times 1{,}000$$

This rate of unmarried females in UK is nearly 4 times that in India where marriage is universal.

Table 15.6: Agewise marital status in 1981

Age groups	15–19	20–24	25–29	30–34	35–39	40–44	45–49
% married	70.1	92.6	94.4	91.7	87.5	78.5	71.2

(Source: FWP in India Year Book 1981–82)

Measures of Morbidity

Morbidity, illness, sickness, or morbid condition mean deviation from a state of physical or mental well-being as a result of disease, injury or impairment. In a given population, morbidity for a given time, may be measured in terms of new cases (incidence) or in terms of new and old cases combined (prevalence).

i. *Incidence Rates*

Incidence tells how often an event occurs in a population over a period of time such as a week, a month, a year, etc., e.g., malaria or typhoid. Accordingly, WHO Expert Committee on Health Statistics, has recommended that the term incidence rate be used to measure frequency of illness that commenced during a defined period, i.e., occurrence of new cases during a specified week, month or calendar year, etc. Spell of sickness means the duration of one illness only. In a defined period, one may have two spells, e.g., two colds in 6 months. The morbidity rates may be calculated for spells or persons and the same may be mentioned in brackets. The denominator for calculation of rates is the mid-period population exposed to risk.

Incidence or inception rate (spells)

$$= \frac{\text{Total number of new spells of illness during a defined period}}{\text{Population exposed to risk in the same period}} \times 1,000$$

Incidence or inception rate (persons)

$$= \frac{\text{Total number of new persons who became ill at least once in a defined period}}{\text{Population exposed to risk in the same period}} \times 1,000$$

ii. Prevalence Rates

Prevalence indicates how common is an event in a population. It is of two kinds:

a. *Period prevalence* It is used to measure the frequency of an illness in existence during a defined period (day, week, month, year, etc). It includes all the cases, in the defined period—old and new cases occurring during the same period.

Period prevalence

$$= \frac{\text{Total number of new and old cases found during a specified period}}{\text{Population exposed to risk at the same period}} \times 1,000$$

b. *Point prevalence* This term is used to measure the number of cases of illness, new and old, existing at a particular point of time, such as at 2 PM on Monday, the 15th July, 1983.

Point prevalence

$$= \frac{\text{Total number of new and old cases found at a particular point of time}}{\text{Population exposed to risk at the same point of time}} \times 1,000$$

Point prevalence is often assessed by survey of a defined population but it may not be possible to survey or cover the entire population at a point of time. Hence, the procedure followed is to continue the survey till completed, from one house to the last not taking into consideration any occurrence reported in the already surveyed population. For example, while finding malaria prevalence rate in a population of 5,000, one may cover only 2,000 persons on the first day. Any occurrence in this population should not be taken into account on the next day, when the remaining population is being surveyed. These are indices of community health but of negative nature, because they measure health by measuring the extent of disease.

Examples

1. In a village with a population 5,000, 20 new cases of cholera occurred in January 1975 and 42 cases in 1976, in the months of June, July and August. Calculate monthly morbidity rate in 1975 and 1976.

Monthly morbidity rate in 1975 = $\frac{20}{5,000} \times 1,000 = 4$, i.e., four cases per thousand per month and

Monthly morbidity rate in 1976 $\frac{42}{3 \times 5,000} \times 1,000 = 2.8/1,000$/month.

2. There were 45 motor accidents in Wardha town with a population of 69,000 in the year 1978. Calculate the incidence rate of accidents.

Incidence rate = $\frac{45}{69,000} \times 1,000 = 0.65/1,000$ per year

Traffic police in Delhi or any large town makes use of this index to assess the benefit of their traffic control measures, year by year.

Prevalence rate, period or point, for spells or persons is calculated in the same way.

3. Tuberculosis survey was done in a town with 15,000 population. It was found that 27 cases were radiologically positive while 6 were sputum positive. Calculate the point prevalence rate of sputum positive cases at the time of survey.

Prevalence rate of sputum positive cases

$$= \frac{6 \times 1,000}{15,000} = 0.4/1,000$$

To determine incidence rate of tuberculosis for any year or month, the fresh cases that occurred in that period, have to be taken into account.

4. If 12 cases of diabetes mellitus were found in 15,000 population in a particular year, calculate the prevalence rate.

$$\text{Prevalence rate} = \frac{12 \times 1,000}{15,000} = 0.8/1,000/\text{year}$$

Morbidity rates may be calculated per 1,000, 10,000 or 1,00,000 depending upon the morbidity load in the community.

Incidence rates are usually calculated for acute disease such as cholera, measles, diphtheria, burns, etc., as their onset is sharply defined. Prevalence rates are determined for chronic illnesses such as leprosy, tuberculosis, filariasis, cancer, diabetes, hypertension, etc. where the onset is very ill-defined. These rates measure the epidemiological situation, suggest priority in measures to be adopted, and assess the progress made in the control of a disease. Prevalence of tuberculosis in 1950s in India was 50 lakhs out of which 5 lakhs recover or die every year but the same number is added up by the incidence of 5 lakhs new cases, thus the point prevalence remains the same in 1990s in spite of the new specific drugs. Reduction in tuberculosis prevalence is a good measure of socio-economic progress or the effectiveness of treatment.

Duration of Illness

It is another measure used for morbidity and is calculated as follows:

1. Days of illness per person exposed

$$= \frac{\text{Total days of illness of all sick persons}}{\text{Total population exposed}}$$

2. Days per illness per ill person

$$= \frac{\text{Total days of illness of all sick persons}}{\text{Number of persons who fell ill}}$$

3. Days of illness per spell

$$= \frac{\text{Total days of illness of all sick persons}}{\text{Number of spells of illness}}$$

Measures of Mortality

i. *Crude Death Rate*

It is the most commonly used vital rate as it is easy to compute. It measures the decline in population due to mortality in general, without specifying the factors that influence the deaths such as age-sex composition of population, occupation, causes of death, etc.

Crude death rate

$$= \frac{\text{Number of deaths which occurred among the population of a given geographic area during a given year}}{\text{Mid–year total population of the given geographic area during the same year}} \times 1{,}000$$

Crude death rate may also be written as:

$CDR = \frac{D}{P} \times 1{,}000$ where D stands for total deaths and P for total population.

Calculation on annual basis is made as per calendar year for international comparisons. In India, it was 13.8, 8.6 and 12.6 in rural, urban and total population, respectively in 1984 (SRS). In 1992, it was 10.9, 7.0 and 10.1 in the same order.

Example

In 1991, the mid-year population of India was 855.7 millions and the total deaths were 83.8516 millions. Calculate the death rate for this year.

$$\text{Death rate} = \frac{83.8516}{855.7{,}00{,}000} \times 1{,}000 = 9.8$$

To study long-term trends over 50 years or so, it is convenient to calculate quinquennial or decennial death rates. Annual rates for periods shorter than a calendar year such as week, month or year, etc., may be computed by multiplying with the **adjustment factors** to compare with those of the corresponding periods of the previous years. Crude death rates are influenced to some extent by age-sex composition of the population and hence should be standardised before drawing conclusions or making comparisons between the places, years, different environmental conditions, etc. Crude death rate in a colony of retired people may be very high, while in a new township as developed on Bhakra dam or an oil refinery, may be very low, being inhabited mostly by young people.

ii. *Standardised Death Rates*

They are found by two methods: direct and indirect.

In the **direct method**, a standard million population is taken, such as of the country as a whole. To this population, age and sexwise death rates of various groups of population whose crude death rate is to be standardised are applied. The weighted average death rate of all age and sex groups of standard million gives the standardised death rate of the place in question. By this method, death rates of any two places can be compared.

In the **indirect method**, death rates of the standard chosen population as of India as a whole, are applied to age and sex groups of the place of which death rate is to be standardised. This gives the index Death Rate. Crude Death Rate of the standardised population divided by the index death rate, gives the Standardising Factor. Then multiply the crude death rate of the place by this Standardising Factor to get the Standardised Death Rate. Indirect method of standardisation can be applied immediately because agewise death rates of standard or chosen population, as of India are readily available for any census year. Detailed calculation is not given. It is needed mostly by demographers and administrators and seldom by medical students or research workers. Methods are simple and can also be used to correct other crude rates as morbidity rate, birth rates, etc. if required.

Table 15.7: Age specific mortality rates—India

Age group (years)	1971	1981	1991	1992
0–4	51.9	41.2	26.5	26.5
5–9	4.7	4.0	2.7	2.9
10–14	2.0	1.7	1.5	1.4
15–19	2.4	2.4	2.1	2.2
20–24	3.6	3.1	2.8	2.8
25–29	3.7	3.2	3.1	2.7
30–34	4.6	4.0	3.1	3.2
35–39	5.7	4.4	3.9	3.8
40–44	6.7	5.8	4.8	5.1
45–49	9.5	8.5	7.4	7.5
50–54	16.8	13.2	11.3	11.5
55–59	21.2	20.6	17.6	17.8
60–64	34.9	33.0	28.5	28.6
65–69	48.4	46.4	41.6	43.8
70+	109.3	97.4	91.4	91.5
All ages	14.9	12.5	9.8	10.1

*Source: Registrar General, India, Selected Socio-Economic Statistics, India, 1996.

iii. *Specific Death Rates*

They are calculated to find the mortality experience in different sections of the population such as infants, mothers, males and industrial workers or mortality due to different causes such as tetanus and tuberculosis. Age-sex death rates are high among the infants and the old and are lowest in the age group 10–14 years (Table 15.7).

a. Age specific death rate

$$= \frac{\text{Number of deaths in the specified age group}}{\text{Mid-year population of the same age group}} \times 1,000$$

It will be more accurate if the deaths in infants are counted by cohort method. In this, all the children born in one calendar year are followed for one year and those who die before completion of the year, are counted. In the year 1992 in the month of June, in India it was 79. It was the lowest in Kerala (17) and highest in Orissa (115). It was above 100 in MP (104) (Source: SRS, July, 1984). The WHO Regulation No.1 classifies deaths in the first year of life—by single day in the first week, by weeks upto the fourth week and by single months up to one year of age.

Death rates under one year of age or infant mortality rate
They have special significance as they are very sensitive indices of health services and socio-economic advancement. Low rates indicate good MCH, obstetric, immunisation and health education services, sanitary environments and adequate nutrition. The rate has gone down from over 200 per 1,000 to below 10 in the 20th century in developed countries while in the developing countries such as India it has declined from 200 to 73 and it falls rapidly with socio-economic advancement.

It was 8.0 in Japan, 13.0 in USA 13.3 in UK, in Argentina 40.8, Sri Lanka 42.4, Thailand 25.5 while in Afghanistan, India and Egypt it was 186, 129 and 84.5, respectively, for the year 1976 (UN Demographic Year Book, 1978–79). The same for India in 1990, was 80 and in 1992, was 79 (J and K census excluded).

Population under one year is difficult to ascertain and is usually under-registered hence the denominator used is total number of live births.

$$\text{Infant mortality rate} = \frac{\text{Number of deaths under one year of age}}{\text{Number of live births}} \times 1,000$$

It will be more accurate if the deaths in infants are counted by cohort method. In this, all the children born in one calendar year are followed for one year and those who die before completion of the year, are counted.

In the year 1985, IMR in India was 97. It was the lowest in Kerala (31) and highest in UP (154). It was above 100 in Assam (111), Bihar (106), MP (122), Orissa (132), and Rajasthan (108) (Source SRS-1985).

The WHO Regulation No. 1 classifies deaths in the first year of life—by single days in the first week: by weeks up to the fourth week and by single months up to one year of age.

a. Foetal death ratio = $\dfrac{\text{Foetal deaths}}{\text{Live births}} \times 100$

Statistics of death below 28 weeks of gestation are usually not available, hence it is better to compute still birth rate instead of foetal death ratio.

Still birth rate or late foetal death rate

$$= \dfrac{\text{Foetal deaths after 28 weeks of gestation}}{\text{Live births + still births}} \times 1,000$$

In India, it was 17.8 in 1972, 15.0 in 1978 and 10.4 in 1984.

b. *Perinatal mortality rate* Still births and early neonatal deaths are attributed to similar causes, hence a single rate called perinatal mortality rate is calculated. It combines:
— foetal deaths after 28 weeks of gestation when the foetus becomes viable,
— deaths during labour, and
— deaths within seven days of neonatal life.

Perinatal mortality rate

$$= \dfrac{\text{Late foetal deaths (after 28 weeks or more) + deaths under one week}}{\text{Total births (live + still)}} \times 1,000$$

In India, it was 50.7 in 1972, 62.2 in 1978 and 53.8 in 1984.

c. *Neonatal mortality rate* = $\dfrac{\text{Number of deaths upto 28 days of life}}{\text{Number of live births}} \times 1,000$

The causes of mortality in this period are antenatal, natal and postnatal. It may be divided into early neonatal up to one week, late neonatal after one week and up to the end of fourth week. In India, it was 71.6 in 1972, 77.4 in 1978 and 65.8 in 1984.

d. Post-neonatal mortality or late infant mortality rate

$$= \dfrac{\text{Deaths after 28 days of life up to one year}}{\text{Live births}} \times 1,000$$

Table 15.8: Year-wise crude death rate and deaths up to one year of age

Year	Crude death rate	Infant mortality rate	Neonatal mortality rate	Post neonatal mortality	Perinatal mortality rate	Still brith rate
1976	15.0	129	77.0	52.0	86.8	17.5
1977	14.7	130	80.2	49.8	63.7	15.5
1978	14.2	127	77.4	49.6	62.2	15.0
1979	12.8	120.0	71.7	48.3	59.0	12.6
1980	12.4	113.9	69.3	44.6	55.7	11.3
1981	11.5	110.4	69.9	40.5	54.6	10.6
1982	11.9	104.8	66.7	38.1	53.2	8.9
1983	11.9	104.9	67.2	37.7	53.6	9.4
1984	12.6	104.0	65.8	38.2	53.8	13.4
1985						

Source: Year Book 1985-86, Government of India, Ministry of Health and Family Welfare, Department of F.W., New Delhi, p. 100 and SRS 1984 and 1985.

The mortality in this period of life is more due to environmental and nutritional factors. It was 68.2 in 1972, 49.6 in 1978 and 38.2 in 1984 in India. Sum of the neonatal and post-neonatal mortality again gives the infant mortality rate. See Table 15.8 for crude DR and DR under 1 year of age from 1976 to 1985 in India.

Example

In a town with a population of 50,000, there were 2,000 births and 200 infant deaths in the year 1966. Of these 80 infants died in the first 28 days of life and 40 of them died in the first week of life. There were 110 still births in the same year. Calculate infant mortality rate, perinatal, neonatal and post-neonatal mortality rates.

Infant mortality rate

$$= \frac{200}{2,000} \times 1,000 = 100$$

Perinatal mortality rate

$$= \frac{110 + 40}{2,000 + 110} \times 1,000 = 71.09 \text{ per thousand}$$

Neonatal mortality rate $\dfrac{80}{2{,}000} \times 1{,}000 = 40$ per thousand

Post-neonatal mortality rate

$$= \dfrac{200 - 80}{2{,}000} \times 1{,}000 = 60 \text{ per thousand}$$

Sex specific death rate Table 15.7 indicates that female mortality rate is higher than that of males till the age of 35 due to neglect and repeated child-bearing. After that it is less than male mortality rate in India. FWP Year Book 1981-82, p. 47).

Male death rate

$$= \dfrac{\text{Male deaths}}{\text{Mid–year population of males}} \times 1{,}000$$

Age-sex specific death rate It gives the rate for specific age and sex e.g.

Death rate of women in reproductive age group 15–49

$$= \dfrac{\text{Female deaths in age group 15–49}}{\text{Female population of age group 15–49}}$$

Maternal mortality rate This rate is important when applied to deaths in mothers owing to puerperal or maternal causes and is specifically known as maternal mortality rate. In this, the population exposed to risk consists of only those women who became pregnant in that calendar year. This number is usually unknown hence the denominator used is live births in that year. Thus,

Maternal mortality rate

$$= \dfrac{\text{Number of deaths due to puerperal causes in females}}{\text{Number of live births}} \times 1{,}000$$

Puerperal or maternal deaths include those due to complications of pregnancy, child-birth and puerperium.

Cause-specific death rate Cause-specific death rates are computed for the total population. Numerator is often too small, hence the rates may be calculated per 1,00,000 instead of per 1,000 to avoid small decimal figures. This way the comparison between two causes becomes easier. Calculations can also be made for specific age, sex or occupation.

Example

In India, having a population of 430 million, 5,00,000 people died of tuberculosis, 50,000 of smallpox, 86,000 of tetanus, 3,00,000 of cancer and 6,10,000 of coronary heart disease in 1961. Calculate the cause specific mortality rates per 1,00,000

Tuberculosis Death Rate

$$= \frac{50,00,000}{43,00,00,000} \times 1,00,000 = 116.28 \text{ per } 1,00,000$$

Smallpox DR

$$= \frac{50,000}{43,00,00,000} \times 1,00,000 = 11.63 \text{ per } 1,00,000$$

Tetanus DR

$$= \frac{8,60,00}{43,00,00,000} \times 1,00,000 = 20 \text{ per } 1,00,000$$

Cancer DR

$$= \frac{3,00,000}{43,00,00,000} \times 1,00,000 = 69.77 \text{ per } 1,00,000$$

Coronary heart disease DR

$$= \frac{6,10,000}{43,00,00,000} \times 1,00,000 = 141.66 \text{ per } 1,00,000$$

Proportional mortality rate In this, death rates due to different causes are compared by expressing them in percentages of the crude or total death rate. Suppose crude death rate is 20 and DR due to tuberculosis is 2, then the tuberculosis DR = 2/20 × 100, i.e., 10% of the total. An increase in the proportional mortality may be due to an increase in the number of deaths from that cause but it can also be due to a decrease in the total number of deaths from all causes. Similarly, decrease in the proportional mortality could be due to an increase in the total number of deaths.

Proportional mortality rate can also be found directly by dividing the number of deaths due to a specific cause by the total deaths and then multiplying with 100. By this, the relative importance of different causes of deaths can be found. An epidemic may upset the picture in some diseases.

iv. *Proportional Mortality Indicator*

It is the ratio of deaths at 50 years and over to total deaths expressed as a percentage. It is one of the measures for level of health, the higher the ratio, the healthier the population.

v. *Case Fatality Rate (CFR)*

It is the ratio of deaths due to a particular disease to the number of persons who suffered from the same disease.

$$\text{CFR} = \frac{\text{Number of deaths due a particular disease}}{\text{Number of cases of the same disease}} \times 100$$

This should not be confused with mortality rates described already.

Example

In a cholera epidemic caused by Asiatic cholera, *Vibrio cholerae*, 64 cases died out of 160 attacks while in typhoid before chloramphenicol, 60 died out of 200. Find the case fatality rates of both and compare.

$$\text{Cholera case fatality rate} = \frac{64 \times 100}{160} = 40\%$$

$$\text{Typhoid case fatality rate} = \frac{60 \times 100}{200} = 30\%$$

Thus, cholera had a fatality higher than typhoid. This rate is also useful to compare the effect of treatment or any other procedure adopted to reduce the mortality such as chloramphenicol treatment in typhoid and preventive inoculation in cholera.

vi. *Life Table*

This is a method of presenting death rates in all the different ages in a selected year. It is meant for finding expectation of life at any age as dealt with in detail in the next Chapter.

Application of Rates and Ratios of Vital Indices

Analysis of health statistics by computing various indices in the form of rates and ratios leads one to draw some conclusions. Interpretation of conclusions should be done carefully to make suitable recommendations to deal with the community health problems. Biased conclusions due to defect in registration, collection and compilation of data are very likely. When the records are not reliable, it is better to carry out sample surveys, following the rules for selection and inquiry strictly.

Monitoring of Family Planning Programme

Birth rates and allied indices are often needed to monitor Family Planning Programme. The common indices in use are:

1. *Crude birth rate*
2. *Fertility rates*—general, married and age-specific.
3. *Child-women ratio*

$$= \frac{\text{Number of children in age group 0–4 years}}{\text{Number of women aged 15–44 years}}$$

It is useful when birth registration is inadequate. It can be calculated from census or sample survey data.

4. *Birth order* Number of children by birth order, after 1, 2, 3, 4, 5, ... goes on falling if programme is going on well.

5. *Spacing* Average interval between any two successive birth orders, reckoned from birth to birth, goes on increasing such as spacing between 1st and 2nd, 2nd and 3rd and so on.

6. *Pregnancy rate*

$$= \frac{\text{Total pregnancies in particular number of married women}}{\text{Total months of exposure in the same women population}} \times 100$$

= number of pregnancies per 1,200 women-months of exposure.

From 100 women between the ages of 30 and 35 years, inquire their dates of marriages and calculate the total number of married years till date. It is presumed that all women in the sample lived with their husbands and all were fertile. Divide the number of pregnancies in these women by the total number of years of their married life.

Example

Out of 100 women questioned, 50 were married 5 years ago and 50 married 10 years ago. They gave history of 630 pregnancies. Calculate the pregnancy rate.

Months of exposure = $50 \times 5 \times 12 + 50 \times 10 \times 12 = 9,000$

Pregnancy rate = $\dfrac{630}{9,000} \times 1,200 = 84$ per 1,200 women-months of exposure

7. *Pregnancy prevalence rate*

a. Pregnancy prevalence rate based on population

$$= \dfrac{\text{Number of women found pregnant}}{\text{Total population}} \times 1,000$$

b. Pregnancy prevalence rate based on women in fertile group

$$= \dfrac{\text{Number of women found pregnant}}{\text{Number of women aged 15–49 years}} \times 1,000$$

Former, though crude, is easier as population of the place is usually known. Survey for finding pregnant women in a given population, say 5000, is done by a nurse or midwife who is familiar with the population. Six weeks amenorrhoea in married women aged 15–49 years can be taken as a workable sign of pregnancy. It can be calculated trimester-wise also. In a population of 10,000 surveyed around Sevagram, it was found to be 24 per 1,000 population in 1974 and 12 in 1977 showing a marked fall after intensive family planning campaign. It rose to 18 in 1978, after the campaign was relaxed.

8. *Percentage of children in age group 0–4* It was 42.2% in 1971 census and the same fell to 39.3% in the 1981 census.

9. *Mean age of women at each parity.* It goes on rising due to delayed marriage, adoption of birth control measures like IUCD, Nirodh, etc., increasing health awareness and changing socio-economic conditions.

10. *Eligible couples per thousand population* It means the number of married couples in a thousand population to be approached for adopting family planning methods. In such couples, female spouse is between 15 and 44 years of age. The number falls due to rising age of girls for marriage and sterilisation drive in that age group.

11. *Target couples per thousand population for sterilisation* It means the number of married couples with two or more children when female spouse is aged 15–44 years. They are required to be motivated for sterilisation operation of either of the partners. Number goes on falling due to increasing number of women already sterilised in the defined population.

12. *Mean parity at the time of sterilisation* It goes on falling in the continued family limitation programme.

13. *Mean age at sterilisation* It goes on falling because younger couples come forward for sterilisation as the family planning drive goes on.

Monitoring of MCH Services

Infant mortality rates are split into perinatal, neonatal and post-neonatal rates to estimate the extent of the problem, to report measures adopted and to assess the progress made in MCH and general health services. For correct interpretation cohort analysis method should be preferred when surveys are done.

Conclusions from high maternal mortality rate are drawn accordingly and measures like improved antenatal, natal, postnatal and family planning services are proposed.

Monitoring of Illness in the Community

Morbidity rates From the incidence or prevalence of diseases, not only is the extent of illnesses in the country found but also conclusions are drawn about the role of environmental conditions. High incidence of cholera and gastrointestinal infections means poor water supply, drainage and disposal of refuse. High prevalence of tuberculosis means poor socio-economic conditions. Certain living conditions contribute to individual and community health. These can be assessed, such as food and nutrition, housing, place of work, recreation, social security, atmospheric pollution, etc. These are indicative of "level of living" and measure whether the desired standards of living have been achieved or not. They do not measure the actual health condition. Thus, negative indices of community health are depended upon to draw conclusions about the health of a community.

Of these indices, one should search for the index that can be the most comprehensive yardstick and then find the specific ones.

Comprehensive Indicators that Measure Health

1. *Proportional mortality indicator* As defined before, it is obtained as a percentage of the deaths at 50 years and above to the total deaths. There are more old people in developed countries and large number of deaths are there at the age of 50 years and over. In countries with poor health conditions and poor health services, mortality is higher in younger children, in mothers due to specific causes and in other due to preventable and controllable communicable diseases. So, higher the percentage value of this index, the better is the health condition of the community. In this index, the population data are not required and slight defects in registration do not affect the conclusion and interpretation.

2. *Expectation of life* Longevity of life worked out by the life table method, is also a comprehensive measure of community health and summarises the mortality experience at all ages of life. It is not affected by age and sex distribution. It is obtainable at 10 yearly intervals in census years only and measures health on long-term basis, so it may not be a suitable guide to adopt any one public health measure. In India, it was 52.09 for both sexes, 52.62 for males and 51.55 for female in the period 1976–81. In all developed countries,

women live longer on an average by 5 years or more. In USA expectation of life at birth was 68.70 in males and 76.50 in females, the same ratio in Thailand was 65.40 and 60.80. Number of old people above 60 years was 5.1% and in 1981 it was 5.6% (DGHS Chronicle).

3. *Number of females* It is always higher than males in socio-economically advanced countries while in the underdeveloped countries, number of males is higher than females due to better health care of males.

4. *Crude death rate* This is usually the index available but should be made use of, only after adjusted or standardised rates are worked out. It is useful for short-term comparisons.

Specific Indicators that Measure Health

They include still births, perinatal, neonatal, post-neonatal, infant and maternal mortality rates.

Death rates from communicable diseases such as tuberculosis, cholera and tetanus are useful indicators of community health, though records are often incomplete and not dependable. Total death rate in the age group 1–4 years is also a useful indicator.

Table 15.9: Targets for "Health for all by the year 2000 AD"

Index	1981	1990	2000 AD
Crude birth rate	33.2	27.0	21.0
Crude death rate	12.5	10.4	9.0
Infant mortality rate	127	80–90	< 60
Perinatal mortality rate	60–109	–	30–35
Maternal mortality rate	5–8	–	> 2
Life expectancy at birth	52.6 for males	58	64
	51.6 for females	57.7	64
Percentage effective			
Couple protection rate	22	44	60
Net reproduction rate	1.67	–	1.0
Natural growth rate	1.9	1.66	1.25
Family size	4.3	–	2.3
Percentage of deliveries by trained birth attendants	10–15	–	100

Source: Health Statistics in India, CBHI, DGHS, Govt. of India, Year 1982. P. 114. Report of the Working Group on Health for all by 2000 AD Ministry of Health and FW

Morbidity data collected in specific surveys can be a good indicator of comprehensive or specific health services. Nutritional status, absenteeism due to sickness in schools and industries and development of children from 1–4 years may be useful guides in the assessment of community health.

Chapter 16

Life Table

Under the measures of mortality described in the previous Chapter, an important measure: "**Expectation of Life**" at birth or at any other age was not discussed. This involves the use of life table which is a particular way of expressing death rates experienced by a particular population during a particular period. Life table is a special type of cohort analysis which takes into account the life history of a **hypothetical group** or **cohort** of people that decreases gradually by death till all members of the group died. It is a simple measure or device not only for mortality but also for other vital events like natality, reproduction, chances of survival, etc.

USES AND APPLICATION

1. To find the number of **survivors** out of 1,000 or 10,000 or over, birth or at any age thereafter say:
    i. At the age of 5 to find the number of children likely to enter primary school.
    ii. At the age of 15 to find the number of women entering fertile period or find the number of adolescents entering the community.
    iii. At the age of 18 to find the number of persons who become eligible for voting; at 55 or 58 or 60 who become eligible for pension; and at 65 and over to assess the geriatric problems and solutions like income tax benefits, Railway concession, etc.
    iv. At the age of 45 or above to find the number of women reaching menopause.
    v. Survival at different ages of women in fertile period and birth of female children from the age 0 to 15 which helps

in finding agewise fertility rate, net reproduction rate, etc.
2. To estimate the number likely to die after joining service till retirement, helping, thereby in budgeting for payment towards risk or pension.
3. To find expectation of life or longevity of life at birth or any other age. Increase in longevity of life means reduction in mortality. Thus, life table is another method applied to compare mortality of two places, periods, professions or groups.
4. To find survival rate after treatment in a chronic disease like tuberculosis and cancer or after cardiac surgery like bypass or heart transplantation by modified life table technique.

In short, life table helps to project population estimates by age and sex. Willinm Farr called it a **biometer of population.**

Table 16.1 indicates the longevity of life or expectation of life at birth by sex in different states of India worked out for year 1986–90.

Table 16.1: Expectation of life at birth by sex during 1986–90

States	Male	Female	Person
Andhra Pradesh	58.2	60.4	59.1
Assam	53.6	54.2	53.6
Bihar	55.7	53.6	54.9
Gujarat	57.0	58.8	57.7
Haryana	62.2	62.2	62.2
Himachal Pradesh	62.4	62.8	62.8
Karnataka	60.4	62.6	61.1
Kerala	**66.8**	**72.3**	**69.5**
Madhya Pradesh	53.7	53.0	53.0
Maharashtra	61.2	63.5	62.6
Orissa	54.6	54.0	54.4
Punjab	64.7	66.9	65.2
Rajasthan	55.2	56.2	55.2
Tamil Nadu	60.0	60.6	60.5
Uttar Pradesh	54.2	52.5	53.4
West Bengal	60.2	61.2	60.8
All India	57.7	58.1	57.7

Source: Office of the Registrar General, India Sample Registration System (SRS)

CONSTRUCTION OF A LIFE TABLE

To construct a life table, two things are required:
1. *Population living at all individual ages in a selected year.*
2. *Number of deaths that occurred in these ages during the selected year.*

Selected year should be the most recent one for which accurate statistics are available.

Basic element of life table is q_x, i.e., the probability of dying from age x_0 to x_1, x_1 to x_2, ... to x_n. It is computed per person for each year or for a span of years. Calculation of death rates is based on the mid-year population. To calculate the mid-year population for first 24 months of life, needs special attention as the deaths do not occur uniformly in this period.

From third year onwards, mid-year population is calculated easily because deaths are more or less evenly distributed. Half the deaths occur by 30th June and the other half after 30th June. So half of deaths can be added to the population of **cohorts** that started life (l_x) at any age to get mid-year population. Person-years lived (L_x) is also easy to calculate. It will be equal to $l_x + 1/2 d_x$.

During infancy and early childhood, mortality changes rapidly with age even within the interval of a single year. Hence, presumption of equal distribution of deaths in two halves of first 2 years is not valid.

To calculate mid-year population following assumptions for two halves of a year are considered useful (x implies age, m stands for months and y for years)—
(in the first year of life)

x_{0-6m} : x_{6m-12m} : : 0.7 : 0.3 (in the first year of life)
$x_{12m-18m}$: x_{18-24m} : : 0.6 : 0.4 (in the second year of life
$x_{2y-2.5y}$: $x_{2.5y-x3y}$: : 0.5 : 0.5 (in the third year onwards)

Computation of mid-year population in any year or span of years as well as of person-years contributed to l_x by d_x, is done by demographers on these assumptions depending upon the infant and toddler mortality and availability of records, etc.

P_x, i.e., probability of survival is equal to $1 - q_x$ because total probability is one.

Imagine a cohort of 1,00,000 newborns, starting life together (Table 16.2). Subject them (l_0) and the survivors at each age (l_x) to the mortality rates of the selected year, till all members of the cohort die.

Table 16.2: Construction of a life table

Age	Number started life	Number died	No. of person-years lived	Total person-years lived	Expectation of life
x	l_x	d_x	L_x	T_x	e_x
1	2	3	4	5	6
0	1,00,000	10,000	92,500	92,500	$\dfrac{\Sigma L_x}{l_x}$
1	90,000	2,700	88,650	181,150	
2	87,300	1,746	86,427	267,577	
3	85,554	1,711	84,699	352,276	
4	83,843	1,258	83,214	435,490	
5	82,585	991	82,090	517,580	
6	81,594... and so on				

Indicate age x by suffixing the number of years at the foot such as $x_0, x_1, \ldots x_n$, survivors l_x at different ages as $l_0, l_1, l_2, \ldots, l_n$, number died d_x as $d_1, d_2, \ldots, d_n$, years lived L_x as $L_1, L_2, \ldots, L_n$ and total years lived as T_x as $T_1, T_2, \ldots, T_n$.

Column x indicates the age at which cohorts start life such as at birth indicated by 0 or at any age after that such as at age 1, age 2, and so on.

Column l_x gives the number that *started life* at any particular age such as 1,00,000 at birth, 90,000 survivors at age 2 and so on.

Column d_x is the number *died* in each year, such as 10,000 in the first year from age 0 to age 2,700 in the second year and 1746 in the third year and so on. This number is found at the specific mortality rate of each year, e.g., in the first year from age 0 to age one, 10,000 died at the infant mortality rate of 100 per 1,000. Deaths in the 2nd, 3rd, 4th, 5th and 6th year of age are calculated in the Table 16.1 at the specific death rates of 30, 20, 20, 15 and 12, respectively.

L_x column gives the estimated total number of person-years *lived* by the cohort at each age. It will always be more than the survivors at the end of any particular year because all do not die in the beginning of the year. They live for some days or months in each year before death, e.g., in the first year, 10,000 died and 90,000 survived (1,00,000 – 10,000). In this year most deaths occur in first week and first month and less and less later, even in the countries with high infant mortality rate such as 100. It may be presumed that 10,000 before death, lived for 3 months on an average so they contributed (3/12 × 10,000) 2500 years. Thus, total years lived L_0 in the first year will be 92,500 (90,000 + 2,500). In subsequent years, it is presumed that a person lives for half a year on an average before death. L_1 or number of years lived up to the end of second year will be 87,300 + 1/2 of 2700 = 88,650 and so on.

Column T_x gives the *total number of years lived till any age*, e.g., 1,00,000 that started life, lived 90,000 + 2500 = 92,500 years up to age one as entered under column opposite age one, same as L_1.

Up to the age two, total number of years lived will be L_1 till the end of first year 92,500 + L_2, i.e., years lived till the end of second year, 8,86,550 = 1,81,150 as entered opposite age two.

Uptill the end of 5 years, $T_5 = L_1 + L_2 + L_3 + L_4 + L_5$ (92,500 + 88,650 + 86,427 + 84,699 + 83,214 = 4,35,490). T_{50} or total years lived up to age 50 will be the total years lived by the starters till 50 years passed ($L_1 + L_2 + L_3$, ..., L_{50}). Final T_n, when all cohorts die off, will give the number of years lived by 1,00,000 that started life (l_0).

The mean expectation or longevity of life, i.e., average number of years a person is likely to live at age 0 or at any age after that is denoted by e_x. This is obtained by subtracting the years already lived from the final total of years lived and dividing the balance by the number of starters at the age, longevity is desired to be calculated.

$$e_x = \frac{\Sigma L_x}{l_x} = \frac{\text{Sum of person – years lived till age x}}{\text{Number of starters}}$$

Mean or average expectation of life (e_x) can also be found at any age (x) from the survivor column l_x by the formula—

$$e_x = \frac{\text{Sum of } l_x \text{ column excluding those starting life}}{\text{Number that started life}} + \frac{1}{2}$$

Sum of l_x gives the total person-years lived. Dividing by the number of starters, gives mean years lived per person. Half year is added as an average period lived after completion of the last year.

Longevity of life can also be calculated from the column of died, d_x, by the formula—

$$e_x = \frac{d_{x0} \times \frac{1}{2} + d_{x1} \times 1\frac{1}{2} + d_{x2} \times 2\frac{1}{2} + \ldots + d_{xn} \times n\frac{1}{2}}{\text{Number that started life}}$$

Number of years lived by the deceased in each age group are found by multiplying the deaths by age + 1/2 years, i.e., $d_x \times (x + 1/2)$. The d_{x0} means the number of deaths that occurred in the first year of life from age 0 to age 1. When multiplied by 1/2, it gives the number of years lived by the deceased in the first year. The d_{x1}, means the number of deaths in the second year. When multiplied by $1\frac{1}{2}$ it gives the number of years lived by the deceased in the second year, and so on. Thus, d_x column can also give the number of years lived by all before death. This is more laborious than the calculation from l_x.

Life tables can separately be prepared for sexes, professions, places and periods and comparisons made on the basis of **life expectancy**. Such comparisons are independent of age and sex composition.

Cohort analysis by life table method can be used as a measure of total fertility. Start with a cohort of women in child-bearing age and subject the different age groups to fertility rates of most recent year for which accurate statistics are available as mentioned on pages 253 and 254. Net reproduction rate (NRR) described on the same pages is also calculated by life table method.

Life Table

Population and deaths for ages 0–4 are given in the Table 16.3 below. Suppose 10,000 children start life, make a life table for them and find the number surviving at each successive birthday till the 5th.

Table 16.3

Age x	Population	Deaths
0–	4151	70
1–	4792	16
2–	4797	6
3–	4998	4
4–5	4798	4

Solution

In the first year of life, all the 70 deaths did not occur on the first day. Probably 7/10th died by mid-year, therefore, mid-year population of first year would have been $4151 + 7/10 \times 70 = 4200$. Then the chance of dying per person i.e., $d_x = \frac{70}{4200} = 0.0167$. Thus, 167 ($d_x$) out of 10,000 would actually die, so the number of survivors entering life for 2nd year is $10,000 - 167 = 9833$ (Table 16.4).

After the first year, we presume that half the deaths occurred before and half after six months. When 16 deaths occurred, they lived 8 years more. These 8 years are to be added to those that reached 2nd year, i.e., $4792 + 8 = 4800$. Chance of dying per person $= 16/4800 = 0.0033$. Multiply 9833 by 0.0033 to find the number died in the second year at this rate, $9833 \times 0.0033 = 32$. Population surviving to start the third year of life will be $9833 - 32 = 9801$.

Calculate the number of deaths out of 9801, and similarly, the number of those starting life in 4th and 5th years of life and find the number that survived to start 6th year or to celebrate 5th birthday.

Table 16.4: Number of children expected to complete 5 years of age out of 10,000, found by constructing a life table as done on page 278

Age	Mid year population, i.e., population + deaths up to 6 months	Probability of dying	Probability of surviving	Population that died out of 10,000 at the rate, qx	Population that started life	Expectation of life
x	l	q_x	$p_x = 1 - q_x$	d_x	x	$e_x = \dfrac{\Sigma l_x}{starters} + \dfrac{1}{2}$
0–1	4151 + 7/10 of 70 = 4200	$\dfrac{70}{4200} = 0.0167$	0.9833	10000 × 0.0167 = 167	10,000	
1–2	4792 + 1/2 of 16 = 4800	$\dfrac{16}{4800} = 0.0033$	0.9967	× 9833 × 0.0033 = 32.45 (32)	10,000−167 = 9833	
2–3	4797 + 1/2 of 6 = 4800	$\dfrac{6}{4800} = 0.0013$	0.9987	9801 × .0013 = 12.74 (13)	9833−32 = 9801	
3–4	4998 + 1/2 of 4 = 5000	$\dfrac{4}{5000} = 0.0008$	0.9992	9788 × 0.0008 = 7.8 (8)	9801−13 = 9788	
4–5	4798 + 1/2 of 4 = 4800	$\dfrac{4}{4800} = 0.0008$	0.9992	9780 × 0.0008 = 7.8 (8)	9788−8 = 9780	

Life table constructed as per example 1, page 281. From 2nd year onwards. For ease in calculations it is assumed that the deaths occurred uniformly throughout the year.

Thus, life table up to the age of 100 years can be constructed and seen when all in the population of 10,000 that started life disappear at the death rates of chosen year.

Example 2

From the abridged SRS based life Table 16.5: India, 1976–77 for males and females, find:
  a. What proportion of men entering service at 20 will be eligible for pension, at the age 55 and at 58?
  b. The expectation of life at birth for males is 0.8 years more than for females. Does this mean that men of 65 have an average about 0.8 years more to live than women of the same age?
  c. Calculate the expectation of life for men at the age 55. What assumption have you made to do this? Does your result help you to answer the question (b)?
  d. If 20% of deaths occurring in men between the ages of 60 and 70 inclusive are due to cancer, what proportion of men aged 60 years are likely to die of cancer before reaching their 70th birthday?

Solution

(a) l_{20} or number of men who entered service at the age 20 = 78,368 and l_{55} or number of men who completed 55 years = 64,886 number eligible for pension $\frac{64,886}{78,368} \times 100 = 82.79\%$

l_{58}, i.e., the number of men who completed 58 years of age = L_{58} − 3/5th of $(l_{60} - l_{50})$ = 64,886 − 0.6 × (64,886 − 58,985)
  = 64886 − 0.6 × 5901 = 64,886 − 3,541 = 61,345

3/5th of the deaths in 5 years, from 55 to 60 have to be deducted from those living at the age 55 to get l_{58}, i.e., the number reaching 58 years of age. So percentage of those who become eligible for pension at the age 58 years

$$= \frac{l_{20}}{l_{58}} \times 100 = \frac{61,345}{78,368} \times 100 = 78.27\%$$

(b) Difference in expectation of life between males and females at birth = 50.8 − 50.0 = 0.8 (longevity greater in men).

Table 16.5: SRS based life table: India, 1976–77

Age interval (exact age) x to x + n	Males				Females			
	nq_x	l_x	nL_x	$e_x°$	nq_x	l_x	nL_x	$e_x°$
0–1	0.12650	1,00.000	90.841	50.8	0.13460	1,00.000	90.255	50.0
1–5	0.07502	87.350	3.34.915	57.2	0.10826	86.540	3.28.196	56.7
5–10	0.01982	80.767	4.00.250	57.6	0.02496	77.170	3.81.386	59.4
10–15	0.01045	79.196	3.94.286	53.8	0.01119	75.244	3.74.222	55.8
15–20	0.01193	78.368	3.88.583	49.3	0.01539	47.402	3.69.355	51.4
20–25	0.01440	77.433	3.84.483	44.8	0.02104	73.257	3.62.588	47.2
25–30	0.01391	76.318	3.79.286	40.5	3.02398	71.716	3.54.639	43.1
30–35	0.01933	75.256	3.73.077	36.0	0.02374	69.996	3.46.042	39.1
35–40	0.02471	73.801	3.64.800	31.6	0.02447	68.335	3.37.778	35.0
40–45	0.04336	71.977	3.52.655	27.4	0.02940	66.663	3.29.009	30.8
45–50	0.05765	68.856	3.35.021	23.5	0.03903	64.837	3.16.365	26.6
50–55	0.09094	64.886	3.10.579	19.8	0.06238	62.306	3.02.412	22.6
55–60	0.12953	58.985	2.76.812	16.5	0.09368	58.420	2.87.245	19.7
60–65	0.19855	51.345	2.32.232	13.6	0.16130	52.790	2.43.634	15.5
65–70	0.24782	41.150	1.80.976	11.3	0.21812	44.275	1.98.092	13.0
70+	1.00000	30.952	2.82.667	9.1	1.00000	34.618	3.76.076	10.9

Source—Sample Registration Bulletin Vol. XIV, No. 2, December, 1980 of Registrar General of India, Ministry of Home Affairs, New Delhi.

Difference at 65 = 13.6 − 15.5 = −1.9 (longevity greater in women).

Difference is reversed though it is not the same at 65 years as it was at birth. So it does not mean that men at 65 years have on an average 0.8 years more to live than the women of the same age.

(c) e_x at 55 years for males

$$= \frac{\Sigma l_x}{l_x} + \frac{1}{2} = \frac{\text{Sum of } l_x \text{ column from } l_x \text{ 55 onwards}}{\text{Number of starters at } l_x \text{ 55 years}} + \frac{1}{2} \text{ of 5}$$

i.e., $\dfrac{\text{Number of years lived by all age groups, After 55} \times \text{group interval, i.e., 5}}{\text{Number of starters, i.e., survivors at age 55}} + 2\dfrac{1}{2}$

We add 2.5 years and not 1/2 year because age interval is 5 years. Some died before 2.5 years and others died after 2.5 years. On an average they lived 2.5 years more. Therefore, e_x at 55 years.

$$\frac{\Sigma l_x}{l_x} + 2.5 = \frac{(58,985 + 51,345 + 41,150 + 30,952) \times 5}{64,886} + 2.5$$

$= 2.81 \times 5 + 2.5 = 14.05 + 2.5 = 16.55$

Assumptions made for these calculations are:
a. Nobody lived after 75 years of age.
b. Deaths were evenly distributed along the 5-year intervals. For females e_x at 55 years

$$= \frac{(58,420 + 52,790 + 44,275 + 34,618) \times 5}{62,306} + 2.5$$

$= 3.05 \times 5 + 2.50 = 15.25 + 2.50 = 17.75$

Yes, the result helps in answering question (b). The females at 55 years do not have the same difference in expectation of life as they had at birth. At 55, it is 1.20 (17.75 − 16.55) against −0.8 (50.0 − 50.8) at birth.

(d) Deaths in males between the ages of 60 and 70 years
= 58,985 − 41,150 = 17,835

Cancer deaths = 20% of 17835 = 3567

Therefore, the proportion of men aged 60 years likely to die of cancer before 70th birthday, will be $\dfrac{3567}{58985} \times 100 = 6.05\%$.

286 Methods in Biostatistics

MODIFIED LIFE TABLE

This method is a modification of the usual life table method for calculating the survival rates after specific treatment or operation or at any point of time after that. A total of 23 patients of tuberculosis (Table 16.6) started getting treatment in a TB clinic. Their number became less due to defaulters in the following months. Out of the 23 who started treatment in January, only 22 reported in February, 13 in March and so on till there were only 9 left in the month of November. Similarly, the number of patients who started treatment in February, March, or in any month till October were followed till November in each case, e.g., 20 cases started treatment in July and only 9 remained in November.

Separate calculation of the follow-up rates for, different months becomes somewhat laborious and sometimes difficult too if the number of months are large and the number of patients in each month is small. If the treatment given in all the months is same and other caracteristics of the patients such as age, sex, stage of the disease, etc., were almost similar, then the data can be amalgamated in the life table form. The great advantage of this method is that the entire data in hand can be utilised at some point of time (Table 16.6).

In this method for computing 10 months *follow-up rate*, data of all the patients, whether they attended for the entire period or not, have been used, while in the direct method, the data of only such patients who had attended for the entire period are used for calculating the *follow-up rate*.

For clarification, the data of Table 16.6 are rearranged in Table 16.7.

Thus, as per Table 16.7, the number at 0th (Zeroth) or starting month was 23 + 9 + 20 + 17 + 23 + 17 + 20 + 22 + 18 + 16 = 185. Out of these 185 patients, 22 + 7 + 12 + 15 + 20 + 12 + 12 + 20 + 16 + 12 = 148 reported for treatment in the next month. Therefore, probability of coming for treatment in 1st month, i.e., one month after starting the treatment comes to 148/185 = 0.80.

Table 16.6: Follow-up results of tuberculosis patients from January 1979 to November 1979

Month of treatment	No. of new patients	Feb	Mar	Apr	May	Jun	Jul	Aug	Sep	Oct	Nov
January	23	22	13	12	12	10	9	9	9	9	9
February	9		7	4	4	4	4	3	2	2	2
March	20			12	12	9	9	8	8	8	7
April	17				15	10	10	10	9	7	6
May	23					20	17	12	11	11	9
June	17						12	12	11	9	8
July	20							12	11	9	9
August	22								20	14	13
September	18									16	15
October	16										12

Table 16.7: Follow-up results of tuberculosis patients from January 1979 to November 1979

Month of treatment	No. of patients	Number of patients who came for follow-up treatment by month									
		1st	2nd	3rd	4th	5th	6th	7th	8th	9th	10th
January	23	22	13	12	12	10	9	9	9	9	9
February	9	7	4	4	4	4	3	2	2	2	
March	20	12	12	9	9	8	8	8	7		
April	17	15	10	10	10	9	7	6			
May	23	20	17	12	11	11	9				
June	17	12	12	11	9	8					
July	20	12	11	9	9						
August	22	20	14	13							
September	18	16	15								
October	16	12									
Total	185	148	108	80	64	50	36	25	18	11	9
Number available for next month follow-up	185	136	93	67	55	42	27	19	11	9	

Out of these 148 patients treated in the 1st month, history of 12 patients was not available, therefore, for calculation of follow-up probability for 2nd month, these 12 patients have to be excluded. So 148 − 12 = 136 patients remained in 2nd month for treatment. Out of these 136 patients 13 + 4 + 12 + 10 + 17 + 12 + 11 + 14 + 15 = 108 came for treatment in 2nd month of follow-up. In this way the probability for 2nd month comes out to 108/136 = 0.79. In the 3rd month follow-up, probability is 80/93 = 0.86. Accordingly, probabilities for subsequent order of months are computed and given in p_x column of Table 16.8.

To find defaulters at any month multiply l_x with d_x such as 1,000 × 0.20 = 200 at one month of follow-up.

l_x at one month of follow-up = 1,000 − 200 = 800
d_x till 2 months follow-up = 800 × 0.21 = 168
l_x at second month end = 800 − 168 = 632
d_x till 3rd month end = 632 × 0.14 = 88 and so on

Starting from 1,000 patients, Table 16.8 shows that after 5 months of treatment only 47.5% (475 out of 1000) remain for further treatment. Of starters, 36% (360 out of 1,000) follow treatment for 10 months. Like this, percentage of persons likely to report in different months after starting treatment can be anticipated.

Now fill up the L_x column, i.e., number of months attended by starters l_x at any month. It will be $l_x + 1/2 d_x$ such as at the end of 1st month 800 + 100 = 900 and at the end of 2nd month 632 + 84 = 716 and so on.

Fill up T_x column, i.e., total number of months attended by starters till the end of follow-up after any month such as — L_x at zeroth or starting month = 5,162 and at 1 month = 4,262.

Expected number of months for which a person is likely to attend at any month e_x, is found by the formula T_x/l_x, e.g., at zeroth month, a term is likely to attend for $\frac{5,162}{1,000}$ = 5.16 months. at the end of 6th month 1,990/408 = 4.88 months and so on.

If some patients get lost due to reasons such as transfer, death, accident or some other cause, they must be excluded from the analysis.

Table 16.8: Follow-up results in the life table form

Month of treatment	Probability of reporting next month	Probability of defaulting	Number available in every month out of 1000	Number of defaulters in every month	Number of months attended by the starters	Total months attended by starters at all ages	Number of months a person is expected to attend at any month
x	p_x	$1-p_x = q_x$	L_x	d_x	L_x	T_x	C_x
0	0.80	0.20	1.000	200	900	5.162	5.16
1	0.79	0.21	800	168	176	4.262	5.33
2	0.86	0.14	632	68	588	3.546	5.61
3	0.96	0.04	544	22	533	2.958	5.43
4	0.91	0.09	522	47	499	2.425	4.64
5	0.86	0.14	475	67	442	1.926	4.05
6	0.93	0.07	408	29	394	1.484	3.63
7	0.95	0.05	379	19	370	1.090	2.87
8	1.00	0.00	360	0	360	720	2.00
9	1.00	0.00	360	0	360	360	1.00
10	—	—	360	—	—	—	—

After tabulating these p_x values in the above table, the q_x value a (probabilities of defaulters) are obtained by subtracting p_x value from l_x. For convenience of calculation and comparison l_x i.e. the number that started treatment is taken as 1000.

Similarly, life table of survivorship after any treatment as of cancer by irradiation or drugs or after operation, such as of cancer cervix or breast, can be prepared and made use of in forecasting probabilities of survival at the beginning or at any subsequent point of time.

More recently survivorship after heart operation, such as bypass, angioplasty, ballooning, stenting, heart transplantation, kidney transplantation, etc. and so also after transplantation of lungs, liver and other organs is enquired in terms of probability of living for one year, two years or more. As World registry or WHO is there for such operations. Life table or even a bar diagram (Fig. 16.1) can be constructed

Fig. 16.1: Female life expectancy. Japanese girl child has the longest longevity at birth

on these basis as large amount of data is available at WHO or World level. A full chapter may be required to write on Kaplan-Meir Survival Analysis with examples. It may be possible to include such chapters in the next edition of the book. Even professional figures would not be available to impress on the full utility of life tables in this edition.

Chapter **17**

Exercises

1. Discuss briefly the application and usefulness of 'Methods in Biostatistics' in different disciplines of medicine, in a general way.
2. Write short notes on the following:
    i. Frequency distribution
    ii. Relative frequency
    iii. Cumulative frequency
    iv. Quantitative data
    v. Qualitative data
3. State which type of diagrams are used for presentation of:
    i. Quantitative data
    ii. Qualitative data
    Illustrate giving familiar examples.
4. The following data give the amount of creatinine in mg per 100 ml in a 24 hour urine specimen for normal males.

1.51	1.65	1.58	1.54	1.65	1.40
1.61	1.08	1.81	1.38	1.56	1.83
1.69	1.22	1.22	1.68	1.47	1.68
1.49	1.80	1.33	1.83	1.50	1.46
1.67	1.60	1.23	1.54	1.73	1.43
2.18	1.46	1.53	1.60	1.59	1.49
1.46	1.72	1.56	1.43	1.69	1.15
1.89	1.47	2.00	1.58	1.37	1.40
1.76	1.62	1.96	1.66	1.51	1.31
2.29	1.58	2.34	1.66	1.71	1.44
1.66	1.36	1.43	1.26	1.47	1.52
1.57	1.33	1.86	1.75	1.57	1.83
1.52	1.66	1.90	1.59	1.47	1.86
1.73	1.55	1.52	1.40	1.86	2.02

   i. Tabulate the data in appropriate class intervals using tally mark system.

ii. Present the data graphically in the form of a histogram and a polygon.

5. Illustrate the following data on sterilisation compiled under the Family Planning Programme in India by bar chart (figures in thousands).

Year	Target	Achieved	Per cent achieved
1969–70	2215	1422	64.2
1970–71	2600	1330	51.2
1971–72	2079	2189	105.2
1972–73	5697	3122	54.8
1973–74	2268	942	41.6
1974–75	2000	1354	67.7
1975–76	2492	2669	107.1
1976–77	4294	3261	192.2
1977–78	3990	949	23.8
1978–79	3965	1484	37.4
1979–80	3049	1778	58.3
1980–81	2896	2053	70.9
1981–82	2896	2792	96.4

(Component bar diagram, write percentage on top of bar)

6. Present the following distribution on live births by 'Order of Birth' (India) by bar chart (component or proportional).

Birth order	Rural Percentage		Urban Percentage	
	1972	1978	1972	1978
1	19.85	23.45	21.60	25.65
2	17.44	20.63	18.22	23.33
3	15.70	17.49	16.19	17.88
4	13.77	13.38	14.45	12.22
5	11.24	9.69	9.52	7.93
6 or more	22.00	15.36	20.02	12.99
Total	100.00	100.00	100.00	100.00

7. Illustrate progress in Medical Termination of Pregnancy in India by a line chart using data tabulated below.

Year	Number	Year	Number
1972–73	20,143	1977–78	247,049
1973–74	41,212	1978–79	317,732
1974–75	105,531	1979–80	360,832
1975–76	214,205	1980–81	388,405
1976–77	278,870	1981–82	426,451

8. Prepare a linear diagram of serum levels of oxytetracycline given by intramuscular and intravenous injections from the following data and draw conclusions.

Time taken	Serum level microgram/ml	
	Intramuscular	Intravenous
After 1 hour	0.2	5.91
After 2 hours	1.5	3.39
After 12 hours	0.9	0.493
After 18 hours	0.4	0.035
After 24 hours	0.1	0.0076

Source: JIMA, Vol 59, July 16, 1972, page 60.

9. Prepare sector diagram for the following table.

Type of leprosy	No. of patients
Tuberculoid	148
Lepromatous	64
Indeterminate	18
Borderline	10
Total	240

10. State briefly, giving reasons the kind of diagrams you can draw to present the following types of statistical data.
    i. Yearwise (1, 2, 3, 4, 5) and sexwise distribution of children less than 5 years of age per family in a large town.
    ii. Number of persons using different types of contraceptives—diaphragm, jelly, nirodh, pills, Saheli, copper T in the same town. Present the data by proportional bar diagram.

11. Give your comments on the following statements.
    i. Bar and pie diagrams serve the same purpose.
    ii. Histogram and bar diagrams are not the same.

12. Name and define the statistical measures of central tendency (centring constants) and variability.

13. Following stillbirth rates per 1000 total births were repeated by 30 towns in 1977.
    27, 28, 40, 32, 30,   36, 25, 29, 30, 29   26, 30, 20, 35, 32
    36, 37, 29, 29, 42   32, 27, 35, 36, 29   33, 27, 41, 49, 34
    i. Prepare a frequency table using class intervals such as 20–24, 25–29, 30–34, ..., etc.

ii. Transform tabular presentation to graphic presentation. Can the above data be presented by line diagram, simple bar diagram or histogram.
14. Calculate mean, median and mode of the data given in exercise 13.
15. Determine the range, mean deviation, variance, standard deviation and coefficient of variation of the data in exercise 13.
16. i. Sex ratio of males and females in India in 1981, was 1000 : 933. Literacy rate in the same year was 40% in males and 20% in females. Calculate general literacy rate.
    ii. If birth rate in the same year was 32/1000 population for males and 28 for females, calculate combined birth rate.
17. Calculate the mean number of living children per woman from the following table.

No. of living children (x)	No. of women (f)
0	42
1	49
2	57
3	40
4	31
5	22
Total	241

18. Calculate mean and standard deviation of the following data on protein content of local sorghum varieties:
    9, 8, 8, 9, 8, 9, 10.
19. From the frequency table prepared under exercise 13 show that the mean calculated from a grouped data does not exactly agree with the mean calculated directly from the total number of observations.
20. Compare variability of systolic blood pressure in children of age group 5–10 years, with that of adults of age group 30–40 years. Their means and SDs were 100 and 8 in children and 120 and 12 in adults, respectively.
21. Define percentiles and discuss their applications. Compute first, second and third quartiles and semi-interquartile range for the following data.

Height in cm	No. of students
145.0–147.4	2
147.5–149.9	4
150.0–152.4	8
152.5–154.9	18
155.0–157.4	30
157.5–159.9	40
160.0–162.4	40
162.5–164.9	28
165.0–167.4	24
167.5–169.9	6

22. Write short notes on:
    i. Variability
    ii. Biological variability
    iii. Standard deviation
    iv. Coefficient of variation.
23. Blood serum cholesterol levels of 10 subjects are as under:
    240, 260, 290, 245, 255, 288, 272, 263, 277, 250. Calculate mean and SD with the help of assumed mean.
24. Mid-arm circumference (in cm) of 25 male children aged 4 months is given below.
    14,11, 11, 10, 12, 13, 10, 14, 11, 11, 10, 12, 12, 13, 13, 11, 14, 12, 12, 12, 13, 12, 12, 13, 12.
    Calculate the mean, median, mode and SD.
25. Oral surgery unit of a dental college performed the following number of operations each month. Find the range and calculate the mean and SD of monthly operations. 15, 18, 25, 40, 25, 18, 25, 21, 30, 33, 25.
26. The following data show the number of children born to 350 women.

No. of children	No. of women
0	171
1	82
2	50
3	25
4	13
5	7
6	2

   i. Calculate the mean number of children born per woman.
   ii. Draw histogram for the number of children born to number of women.

27. Total serum proteins (in gm per cent) of 24 subjects are as under:
    7.8, 7.2, 7.0, 6.8, 7.4, 7.2, 7.2, 7.4, 7.2, 6.6, 7.1, 7.3, 7.5, 7.4, 7.4, 7.2, 7.2, 6.6, 7.1, 7.3, 7.5, 7.4, 7.2, 7.2, 6.9, 5.8, 7.2, 7.3, 7.0, 7.3, 6.8. Determine the range, mean and SD.
28. Calculate the mean, and coefficient of variation of the following observations on weight in kg of children aged 18 to 21 months.
    8, 5, 10, 10, 9   6, 8, 8, 10, 8   6, 8, 10, 10, 7
    9, 9, 8, 10, 9   10, 5, 9, 8, 9   8, 6, 9, 6, 9
29. Write short notes on:
    i. Normal distribution
    ii. Normal curve
    iii. Random sample
    iv. Cluster sample
    v. Double blind trial
    vi. Biased sample
30. Explain the terms:
    i. Standard error of mean
    ii. Confidence limits
    iii. Null hypothesis
    iv. Level of significance
31. Calculate standard error of mean for data given in exercise 13.
32. Discuss estimation of population parameters from a sample with special reference to mean.
    A random sample of 900 children was found to have a mean fatfold thickness at triceps of 3.4 mm with an SD of 2.3 mm. Can it be reasonably regarded as a representative sample of population having a mean thickness of 3.2 mm?
33. Ten individuals are chosen at random from a normal population and their weights in kg are found to be—68, 63, 66, 69, 63, 67, 70, 70, 71 and 71.
    Does this sample adequately represent a population in which the mean weight was found to be 66 kgs? Justify your answer.
34. Define standard error of mean. Mean Hb% of the population is 14.3. Can a sample of 15 individuals with a mean of 13.5% and SD 1.5 be from the same population?

35. In a universe, the average number of attacks of common cold was 4.5 per person with a standard deviation of 2.87. A sample of 20 persons drawn from the same universe showed the number of attacks of cold per person as follows:

 7, 2, 5, 0, 1   5, 5, 6, 7, 6   2, 6, 9, 4, 8   8, 6, 7, 9, 3

   i. Calculate the mean, SD and SE of mean.
   ii. Comment why SD and SE of mean are different?
   iii. Give 95% confidence limits of attacks of common cold in universe and examine whether all these observations in the sample lie within the confidence limits or not.

36. Determine if height differs with sex.

Sex	Number n	Mean height in cm	SD
Boys	169	168	14
Girls	54	153	8

37. Serum protein is lower in females than in males. Justify this conclusion by applying appropriate statistical technique to the data given below:

Sex	Number n	Mean serum protein level in gm/per 100 ml	SD
Males	18	7.21	0.26
Females	7	6.90	1.28

38. The mean difference in systolic blood pressure of 100 patients was 30 mm of Hg before and after treatment. Standard deviation of difference was 10 mm of Hg. Prove statistically if the treatment was effective.

39. In a group of 196 adults in the age group 45 to 53 years belonging to social class I, the mean serum cholesterol was 180 mg% with a standard deviation of 42 mg%. In a comparable group of 144 adults belonging to social class V, the mean serum cholesterol was 150 mg% with a standard deviation of 48 mg%. Is the difference in cholesterol level of the two classes statistically significant?

40. i. What do you understand by 'probability'? How is it calculated in parametric tests such as standard error of mean?
    ii. Describe the condition which governs the addition and multiplication laws of probability.
41. Prevalence of diabetes in a community is one diabetic case in 10 persons. If 4 children are born what will be the probability of occurrence of diabetes in the following combinations:
    i. All the 4 normal
    ii. 1 diabetic and 3 normal
    iii. 2 diabetic and 2 normal
    iv. 3 diabetic and one normal
    v. All the 4 children born diabetic
42. In two groups of infants in the 8th month of age the following values were observed:

Group	Number of infants	Mean weight	Standard deviation of weights
1	100	6.9 kg	1.10 kg
2	169	7.3 kg	0.91 kg

Test whether mean birth weights are significantly different.

43. In an investigation on neonatal blood pressure in relation to maturity following results were obtained:

Babies 9-days-old	Number	Mean systolic	Standard deviation
Normal	54	75	6
Neonatal asphyxia	14	69	5

Is the difference in mean systolic BP between the two groups statistically significant?

44. The mean plasma potassium level for 50 adult males with a certain disease was found to be 3.35 mEq/litre and the standard deviation was 0.50 mEq/litre. The normal adult value for plasma potassium is 4.6 mEq/litre. Based on the above data, can it be concluded that

males with the disease have lower plasma potassium levels than normal males?

45. Define 't' test, give different situations in which the unpaired and paired 't' tests are applied.
46. Chest circumference in cm of 10 normal children and 10 malnourished children aged one year are given below.
Normal group: 42, 46, 50, 48, 50, 52, 41, 49, 51 and 56
Malnourished group: 38, 41, 36, 35, 30, 42, 31, 29, 31 and 35.
Test for the statistical significance of the difference in chest circumference, on an average between these two groups.
47. Discuss the standard error of mean as a parameter to assess variability.
The following data give the uterine weights (in mg) of 20 rats selected at random from a large stock. Is it likely that the mean weight of the whole lot could be regarded as 24 mg, if the uterine weights follow 't' distribution?

9	18	21	26
14	18	22	27
15	19	22	21
15	19	24	30
16	20	24	32

't' value for 19 df is 2.093 at 5% level.

48. A group of seven patients of rheumatic heart disease with distention of abdomen due to ascites, affecting breathing capacity were treated. Could the change noted in the breathing capacity given in the following table be attributed to treatment?

Maximum breathing capacity (L/min for 7 patients)	Serial No. of patients						
	1	2	3	4	5	6	7
Before treatment	102	89	32	82	36	56	79
After treatment	132	116	50	82	61	64	92

49. In a study of the importance of early and late clamping of the umbilical cord on the blood volume of the infant, the following data was obtained. The total blood volume has been expressed as a percentage of the weight of the baby. Data of 11 babies in whom the cord was clamped early:

    13.8, 8.0, 8.4, 8.8, 9.6, 9.8, 8.2, 10.3, 8.5, 11.5, 8.2

    Data of 16 babies in whom the cord was clamped late:
    10.4, 13.1, 11.4, 9.0, 11.9, 11.0, 16.2, 14.0, 8.2, 13.0, 8.8, 14.9, 12.2, 11.2, 13.9, 13.4

    Does these samples give evidence that there is relationship between the time when the cord was cut and blood volume when expressed as a percentage of body weight?

50. The soporific effect of drugs A and B was studied on ten patients separately. The results were assessed for the additional hours of sleep produced by the drugs. Compare soporific effects of the drugs from the following data:

Additional hours of sleep	Serial No. of patients									
	1	2	3	4	5	6	7	8	9	10
Drug A	0.7	1.6	0.2	1.2	0.1	3.4	3.7	0.8	0.0	2.0
Drug B	1.9	0.8	1.1	0.1	0.1	4.4	5.5	1.6	4.6	3.6

51. Two new types of rations are fed to guinea-pigs. A sample of 12 pigs is fed type A ration and another sample of 12 pigs is fed type B. The gain in weight is given in gm in each case. Compare the weight gain after feeding with the two types of ration.

    Type A   31 34 34 29 26 32 35 38 34 30 29 32
    Type B   26 24 28 29 30 29 32 26 35 29 32 28

52. From the following observations confirm whether the blood glucose level of pigeons when they are exposed to 55°C for an hour goes up significantly when compared with that of control group:

S. No.	Blood glucose level in mg/100 ml	
	Control group	Experimental group
1	200	173
2	186	249
3	176	188
4	184	215
5	170	230
6	172	196
7	170	184
8	163	186
9	176	189
10	173	200

53. Blood glucose level of pigeons is known to be higher than that of rabbits. Prove it by applying proper statistical test to the following data:

S.No.	Blood glucose level per 100 ml	
	Pigeons	Rabbits
1	200	145
2	186	125
3	176	100
4	184	112
5	170	127
6	172	139
7	170	151
8	163	140
9	176	159
10	173	132

NB—Blood was collected after ether anaesthesia in case of rabbits.

54. A group of 15 normal children in a study, had a mean serum iron level of 148 µg% and SD of 44.03. Another group of 15 children with infantile cirrhosis of liver had mean serum iron level of 151 µg% and SD of 49.04. Is the difference between the two serum means statistically significant?

55. A group of 15 normal children in a study had a mean bilirubin level of 1.05 µg% and SD of 0.34. Another group of 15 children with infantile cirrhosis of liver had mean bilirubin level of 4.99 µg% and SD of 2.52. Is the difference between the two means statistically significant?

56. The effects of abdominal distention due to ascites on lung volume and ventilatory function were studied in 15 patients before treatment and after marked reduction of ascites, during a regime of low sodium diet and mercurial diuretics over a ten days' period. Is the difference in maximum breathing capacity following reduction of ascites by treatment significant, if the calculated value of 't' is 1.82?

57. Thirty micrograms of vitamin B_{12} were given intramuscularly every fourth week to six patients of pernicious anaemia during periods of remission. The results are given below. Do the data indicate real improvement in haemoglobin level?

Individual Number	Haemoglobin gm%	
	Before therapy	After 3 months therapy
1	12.2	13.0
2	11.3	13.4
3	14.7	16.0
4	11.4	13.6
5	11.5	14.0
6	12.7	13.8

58. Explain the term 'Null Hypothesis'. Enumerate various tests of significance which prove or disprove null hypothesis.

    In a clinical trial to assess the value of new tranquilliser on psychoneurotic patients with each patient being given a week's treatment with the drug, the drug was considered effective if it lowered anxiety score after treatment. Test the efficacy of drug on the following results.

    Before treatment   22, 18, 17, 19, 22, 12, 14, 11, 19, 7
    After treatment   19, 11, 14, 17, 23, 11, 15, 19, 11, 8

59. The systolic blood pressure of six hypertensive patients were 179, 190, 183, 165, 180 and 175 mm of Hg. After administration of a particular drug for one week the pressures were 175, 180, 187, 150, 170 and 180 mm of Hg respectively. Could such differences arise due to chance?

60. Serum digoxin levels were determined for nine healthy males aged 20–45 years following rapid intravenous injection of the drug. The measurements were made 4 hours after the injection and again at the end of an 8 hour period.

Subject No.	Serum digoxin concentration μg/ml after	
	4 hours	8 hours
1	1.0	1.0
2	1.3	1.3
3	0.9	0.7
4	1.0	1.0
5	1.0	0.9
6	0.9	0.8
7	1.3	1.2
8	1.1	1.0
9	1.0	1.0

Is the difference in the serum digoxin concentration at the end of 4 hours and at the end of 8 hours statistically significant if so at what level?

61. In order to determine the effect of certain oral contraceptive on weight gain, nine healthy females were weighed prior to the start of its use and again at the end of a 3-months period.

Subject No.	Initial weight in kgs	Weight after 3 months
1	48.0	49.2
2	56.4	57.2
3	52.0	56.0
4	60.0	58.0
5	54.0	56.0
6	56.0	57.2
7	48.0	47.2
8	56.0	56.4
9	52.0	52.8

Is there a sufficient evidence to conclude that females experienced gain in weight following 3 months of the oral contraceptive use?

62. Twenty-two patients were divided in two groups A and B. A drug was tried on group A while group B was considered as control. The observations were as follows:

S. No. of patients	Group A Before treatment	Group A After treatment	Group B Control group
1	10	06	03
2	13	09	09
3	06	03	08
4	11	10	06
5	10	10	05
6	07	04	05
7	08	02	07
8	08	05	03
9	05	03	10
10	09	05	08
11	–	–	10
12	–	–	04

For the above observations:
i. Test whether Group A and Group B patients are from the same universe.
ii. Determine whether the drug administered to group A patients is effective.

63. Define standard error of proportion. In a patient of eosinophilia blood report showed 10% eosinophils on examining 200 WBCs. What is the lowest and the highest percentage of eosinophils expected on repeating the differential count?

64. From a universe 40 females using oral contraceptives and 60 females using other contraceptive devices were randomly selected and the number of hypertensive cases from both the groups were recorded as given below:

Type of contraceptive	Total	No. found hypertensive
Oral	40	8
Other	60	15

Test the hypothesis that the proportion of patients with hypertension is the same for the two groups.

65. In a screw manufacturing industry, out of 120 untrained workers 36 were injured during work while of 80 trained workers 8 were injured in the same period of time. Justify the role of training.

66. Define standard error of difference between two proportions. In a survey 1000 males and 1000 females were examined. 234 males and 266 females showed evidence of caries. Test if the difference observed in the two series is statistically significant.
67. A survey of 400 children in age group 0–5 years showed prevalence rate of protein calorie malnutrition to be 15%. Another study showed a prevalence of 5% in a sample of 300 of similar age group. Can we say that there is statistical significance in the difference between the two prevalence rates?
68. In a locality with 1000 unprotected populations, 8 per cent died of smallpox in a specified year. Of the unprotected 250 were vaccinated and only 12 of them died in the following year. The vaccinator claimed that vaccination was responsible for reducing the mortality in the vaccinated population. Justify his claim.
69. In two villages with a population of 5000 each, infants in one village were vaccinated against polio while in the other they were not. In the following year 20 persons died of polio in the unprotected village while only 4 died in the protected one. Could this difference be attributed to vaccination?
70. Define standard error of proportion.

    Occurrence of pyorrhoea among 100 persons who did not brush their teeth was found to be 10, while in another sample of 100 persons who brushed their teeth, the occurrence was found to be 2. Is the difference statistically significant?
71. In an epidemiological study of diabetes in urban and rural populations of Ahmedabad district, the following data was obtained. Compute the prevalence in the areas and determine if the results differ statistically.

Area	Diabetes	No diabetes	Total
Rural	45	3450	3495
Urban	107	3409	3516
Total	152	6859	7011

72. i. Define the term standard error of difference between two proportions. Give examples of its application.
    ii. A total of 200 patients suffering from toothache were divided into two groups of 120 and 80. The first group received drug A and the second group received drug B. The efficacy was measured by reduction in pain after the drug intake. Drug A failed in 36 cases and drug B in 16 cases. What is the SE $(p_1 - p_2)$ between failure rates and what inference can be drawn?

73. In a large maternity home, 270 male and 230 female children were born in one month. Is the sexwise difference in proportions statistically significant if the expected birth proportions of male and female children is 50:50?

74. Test whether the prevalence of carriers of filaria is associated with sex.

Sex	No. of carriers	No. of non-carriers	Total studied
Male	78	412	490
Females	57	553	610
Total	135	965	1100

75. Determine if there is any association between scabies amongst the school children and the socio-economic status of their parents.

Scabies status	\multicolumn{5}{c}{Socio-economic status}	Total
	I	II	III	IV	V	
No. of children with scabies	23	127	640	806	63	1659
No. of children without scabies	427	1573	7560	7526	541	17627
Total	450	1700	8200	8332	604	19286

76. In a study on the effect of dental hygiene instructions on caries in children, 100 children were randomly

selected, 50 were those who received instructions and 50 who did not. At the end of a 6-month period, the children were examined and the number of new cavities found in each child was recorded. The data on new cavities in two groups is given below.

Group	Number of new cavities			Total
	0–1	2–3	4–5	
No. who received instruction	30	15	5	50
No. who did not receive instruction	20	15	15	50
Total	50	30	20	100

Using the 5% level of significance, can it be stated that there is an association between the instructions received in dental hygiene and the number of new cavities?

77. From the data given in the following table, test whether the prevalence of scabies, in two different sexes is significantly different.

Sex	No. with scabies	No. without scabies	Total
Male	1173	10411	11584
Female	547	7644	8191
Total	1720	18055	19775

78. From the table given below, can you conclude that there is an association between the socio-economic status of women and the period of breast feeding.

Socio-economic status	No. of women as per duration of breast feeding in months			Total
	1–6	7–12	12+	
I, II	1	16	2	19
III	1	18	34	53
IV, V	2	11	25	38
Total	4	45	61	110

79. Does the following data suggest any association between the educational status of mothers and their belief about the child's top feed needs?

Top feedings belief	Education status of mothers					Total
	Illiterate	Primary school	Middle school	High school	University	
Yes	2	24	25	34	10	96
No	3	6	4	2	0	15
Total	5	30	29	36	10	110

80. Does the data provided below indicate any association between literacy and still births?

Births	Educational status of mothers					Total
	Illiterate	Primary school	Middle school	High school	University	
Live births	12	125	97	92	18	344
Still births	2	21	5	6	1	35
Total	14	146	102	98	19	379

81. In order to examine the effects of tranquillisers and stimulants on driving skills, 150 people selected at random were given either a stimulant, a tranquilliser or an identically appearing placebo. After receiving the medication, the participants were administered a battery of coordination and reaction time tests and the number of "mistakes" for each was recorded. The following table gives the total number of mistakes for the entire battery of tests for the three groups.

Medication	Total number of mistakes			Total
	0–5	6–10	11–15	
Stimulant	10	20	20	50
Tranquilliser	5	15	30	50
Placebo	25	15	10	50
Total	40	50	60	150

Based on this data, can it be concluded that the proportion of mistakes is different in three groups?

82. i. What are the applications of χ^2 test?
    ii. The offspring of 12 families were studied in respect of albinotic children. According to their phenotype 40 of these children were distributed as follows:

Phenotype	No. of children
aa	2
ab	16
bb	22
Total	40

Does the distribution differ significantly from an expected distribution as 1 : 8 : 16 in the same order.

83. In an ophthalmic OPD 170 persons above 40 years were examined. 40 had both trachoma and corneal degeneration while 34 had none. Total cases of corneal degeneration obtained were 101. Determine if there is any association between trachoma and corneal degeneration.

	Trachoma	No trachoma	Total
Corneal degeneration	40	61	101
No corneal degeneration	35	34	69
Total	75	95	170

84. 200 persons above the age of 40 years were examined at an ophthalmic OPD for corneal degeneration. Corneal degeneration was observed in 48 out of 110 persons in the age group 41–50 years; in 30 out of 52 persons of 51–60 years and 23 out of 38 in over 60 years. Determine whether the age plays any role in corneal degeneration.

Age group in years	Corneal degeneration	No corneal degeneration	Total
41–50	48	62	110
51–60	30	22	52
Over 60	23	15	38
Total	101	99	200

85.

Site of infarction	No. of patients with bradycardia	No. of patients without bradycardia	Total
Posterior	31	35	66
Anterior	6	28	34
Total	37	63	100

Test whether the incidence of bradycardia has any predilection for the site of infarction. (JIMA Vol 57, Dec 1, 1971, p. 426).

86. From the following data determine if there is any statistically significant association between parity and type of family:

Parity	Nuclear family		Joint family		Total
	O	E	O	E	
Primipara	12	23.19	35	23.81	47
2nd para	13	19.24	26	19.76	39
3rd para	20	22.69	26	23.31	46
4th para	16	17.76	20	18.24	36
5th para	18	15.29	13	15.71	31
6th para and above	69	49.83	32	51.17	101
Total	148		152		300

87. During a nutritional survey in a random sample of 940 people, nutritional status was found to be as below.

Excellent	140
Good	180
Fair	260
Poor	360
Total	940

Does this proportion differ significantly from an earlier study wherein it was found as 1, 2, 3 and 4 in the same order?

88. In an obstetrical study, 790 expectant mothers of 30 years of age were observed. Of these 480 were primigravida with 30 of them having toxaemia while only 12 of the remaining had toxaemia. Is there any association between toxaemia and gravida number?

89. Test the relationship between the opinion on abortion and duration of married life from the table below.

Duration of years	+ve response		−ve response		Total
	O	E	O	E	
0–4	3	10.27	45	37.73	48
5–8	3	7.49	32	27.51	35
9–12	12	10.06	35	36.94	47
13–16	10	8.56	30	31.44	40
17–20	18	10.92	33	49.08	51
21 and above	12	10.70	38	39.30	50
Total	58		213		271

IJ of P and SM, No. 4 Vol I, June 1970 p. 205

90. 72 died out of 1200 patients of cholera treated on Regimen A while 45 died out of 945 patients on Regimen B. Which of the two regimens is statistically better?

91. A random sample of adults was examined for the presence of guineaworm infestations prior to health education on sanitation. 175 out of 2500 were found to have infestation. A second random sample was examined a few months after the health education programme, when 192 out of 3250 adults were found to have the infestations. What do these figures suggest?

92. The following table gives the results of neomycin prophylaxis in cross-infection of burns during first 14 days.

Dressing with	Staphylococcal infection		Total
	Acquired	Not acquired	
Penicillin cream plus neomycin	5	28	33
Penicillin cream alone	18	12	30
Total	23	40	63

What conclusions will you draw from this data?

93. A total of 170 children with a diagnosis of pertussis were grouped at random in 3 groups for treatment by different methods. The results of treatment at the end of six days for the different types of therapy are given below. Do these data give any evidence of difference in the two therapeutic groups?

Treatment	Total number of Patients	Number of Failures
Control	55	23
Streptomycin	66	28
Chloramphenicol	49	17
Total	170	68

94. What is the utilitiy of a scatter diagram? Calculate correlation and regression coefficients to show association between the following two variables.

Variable	Serial number of subjects									
	1	2	3	4	5	6	7	8	9	10
X	2.0	3.0	4.5	5.0	6.0	7.0	7.5	8.5	9.5	10.0
Y	1.5	2.0	3.5	3.5	3.5	4.0	3.0	4.5	4.5	5.0

95. During a laboratory experiment muscular contractions of a frog muscle were measured against different doses of a given drug. The height of the curve was considered as the respone to the drug. The observations were as below.

	Serial number of experiment				
	1	2	3	4	5
Dose of drug	0.3	0.4	0.6	0.8	0.9
Response to drug	54.0	59.0	60.0	65.0	70.0

For the above data:
a. Calculate correlation coefficient and its significance.
b. Determine the regression coefficient $b_y x$.
c. Determine the expected values of Y for the given values of X using regression equation $Y = a + bx$.

96. a. What is the utility of the knowledge of correlation coefficient in the field of dentistry?
b. For the following data, calculate correlation coefficient to determine association if any, between fluoride content of drinking water and community fluorosis index:

Fluoride level in drinking water (in mg/litre)	Community fluorosis index (in percentage)
0.8	0.1
1.3	0.4
1.5	0.8
1.9	0.6
2.3	0.7
2.3	1.1
2.4	0.8
2.6	1.1
3.5	1.6
3.6	1.3

(The correlation coefficient value for 8 df at 0.05 level r = 0.632, 0.01 level r = 0.765; and 0.001 level, r = 0.872).

97. What are the sources for collection of vital statistics? Discuss their utility in practice of Community Medicine and Public Health Administration.

98. The census population of a certain city on March 1, 1971 was 338,750 and on March 1, 1981 it was 402,565. Calculate the mid year (i.e., July 1) expected population for the year 1973 and 1984, by AP method.

99. How do ratio, proportion and rate differ in expressing vital indices? Give familiar examples.

100. Differentiate:
    a. Crude and specific vital rates,
    b. Birth and fertility rates,
    c. Incidence and prevalence rates,
    d. Crude and standardised death rates,
    e. Perinatal and neonatal mortality rates.

101. The mid year population of a city was 20,52,000. In the same year, the number of attacks from cholera was 228 and the number of deaths was 57. Calculate the morbidity rate due to cholera and the case fatality rate.

102. Calculate the crude birth rate, crude death rate and infant mortality rate of a town whose mid year population is 5,00,000 live births in a year were 15,000, number of deaths 7,000 and the number of infant deaths 1875.

103. In a town with a mid-year population of 2,00,000 there were 6050 live births, 110 still births, 250 deaths

within one week after birth and 480 deaths in the first month of life and 750 deaths in the first year of life, in a particular year. Calculate still birth, perinatal and infant mortality rates and compare with national rates in 1985.
104. Explain indices used to monitor progress in Family Planning Programme.
105. Calculate the number of children produced in the year 1978 per woman per year at the GFRs given in Table 15.4.
106. From the SRS Table 16.4 calculate what percentage of:
    a. children reach school age, i.e., 5 years of age
    b. females reach menarche at 15 years of age
    c. females become marriageable at 18 years of age
    d. males become marriageable at 21 years of age
    e. women reach menopause at 45 years of age
107. What indices are used to measure level of health. Compare the health level in India reached in the year 1951 and 1981. What level is expected to be reached by the year 2000 AD?
108. Is there any significant association between number of sex partner and positivity of HIV from the following data?

No. of sex partner	HIV +ve	HIV −ve	Total
Multiple	12	252	264
Single	1	74	75
Total	13	326	339

109. Discuss the associability of a history of STD and positivity of HIV from the given information.

History of STD	HIV +ve	HIV −ve	Total
Yes	1	23	24
No	13	408	421
Not known	26	338	364
Other	2	200	202
Total	42	969	1011

110. Does any sex partner play any role for the positivity of HIV?

Sex partner	HIV		Total
	+ve	–ve	
Male	0	8	8
Female	13	311	324
Both	1	6	7
Total	14	325	339

Answers to Exercises

14. $\bar{X}$ = 32.33, Median = 31.69, Mode = 28.35
15. Range = 20–49, Mean deviation = 4.6, Var = 34.09, SD = 5.89, CV = 18.31%
16. a. GLR = 30.346   b. CBR = 30.06
17. 2.461 per woman
18. $\bar{X}$ = 8.71, SD = 0.760
19. $\bar{X}$ from grouped data = 32.3, from n observations = 32.167
20. Adults 10%, children 8%
21. q_1 = 156.5, q_2 = 159.88, q_3 = 165.52, SIQR = 4.51
23. $\bar{X}$ = 264, SD = 17.372
24. $\bar{X}$ = 12, Median = 12, Mode = 12, SD = 1.17
25. Range 15–40, $\bar{X}$ = 25, SD = 6.928
26. $\bar{X}$ or Mean = 1.5
27. Range 5.8–7.2, $\bar{X}$ = 7.13, SD = 0.375
28. $\bar{X}$ = 8.23, SD = 1.55, CV = 18.83
31. SE = 1.1
32. No, Z = 2.63, p < 0.01
33. Yes, 66 is within 2 SE, t = 1.86 SE = 0.95
34. Yes, t = 2.05, SE = 0.39, p > 0.05
35. a. $\bar{x}$ = 5.3, SD = 2.59, SE = 0.58 at 95% confidence limits.
    b. SD is variabtion within the sample while SE is variation from sample to sample.
    c. Conf. limits 6.46 and 4.14

Exercises 317

36. $Z = 9.8$, highly significant, $p < 0.001$
37. $t = 0.64$, insignificant, $p > 0.10$
38. Highly effective, $Z = 10$, $p < 0.001$
39. $Z = 6.0$, significant $p < 0.01$
41. a. 0.6561   b. 0.2916   c. 0.0486   d. 0.0036
    e. 0.0001
42. $Z = 3.07$, significant, $p < 0.01$
43. Yes, $t = 3.44$, $p < 0.001$
44. Yes, $Z = 17.8$, $p < 0.001$
46. $SE = 2.03$, $t = 6.73$, $p < 0.001$
47. No, $t = 2.72$, $SE = 1.25$
48. Yes, $t = 4.18$, significant, $p < 0.01$
49. Yes, $t = 2.57$, significant, $p < 0.02$
50. Insignificant, $t = 1.99$, $SD = 1.6$, $SE = 0.50$, $p > 0.05$
51. Sig., $t = 2.34$, $p < 0.05$
52. Sig., $t = 2.95$, $p < 0.01$
53. Sig., $t = 9.50$, $p < 0.001$
54. Insig, $t = 0.17$, $p > 0.10$
55. Yes, $t = 6.00$, $p < 0.001$
56. No, $t = 1.82$, $p > 0.05$
57. Highly sig., $t = 5.93$, $p < 0.01$
58. Insig., $t = 1.06$, $p > 0.05$
59. Yes, Insig., $t = 1.5$, $p > 0.10$
60. Sig., $t = 3.0$, $p < 0.05$
61. $t = 1.26$, $p > 0.10$
62. a. $t = 1.77$, $p > 0.05$ b. $t = 5.58$, $p < 0.001$
63. 5.8% and 14.2% (95% conf. limits)
64. $Z = 0.59$, Insig., $p > 0.10$, $\chi_1^2 = 0.12$ with Yates' correction
65. $SEp = 5.36$, $Z = 3.73$, $p < 0.001$
66. $SE = 1.934$, $Z = 1.653$, Insig., $\chi_1^2 = 2.56$
67. $SE = 2.184$, $Z = 4.58$, $\chi_1^2 = 16.89$ $p < 0.001$
68. $SE = 1.6$, $Z = 2$, $\chi_1^2 = 2.55$
69. $SE = 0.028$, $Z = 3.265$, $\chi_1^2 = 9.40$
70. $SE = 3.31$, $Z = 2.417$, $\chi_1^2 = 4.34$
71. $SE = 0.347$, $Z = 5.04$, $\chi_1^2 = 24.65$ $p < 0.01$

72. SE = 6.12, Z = 1.63, χ_1^2 = 2.44 Insig.
73. χ_1^2 = 3.2, p > 0.05
74. χ_1^2 = 11.08, p < 0.01, 10.91 with Yates correction
75. χ_1^2 = 30.90, < 0.00
76. χ_1^2 = 6.0, p < 0.05
77. χ_1^2 = 71.40, p < 0.001
78. χ_4^2 = 19.89, p < 0.001
79. χ_4^2 = 9.93, p < 0.001
80. χ_4^2 = 8.82, p > 0.05
81. χ_4^2 = 27.25, p < 0.001
82. χ_2^2 = 1.41, p > 0.05
83. χ_1^2 = 1.63, p > 0.05
84. χ_2^2 = 4.677, p > 0.05
85. With Yates' correction χ_2^2 = 7.07, p < 0.05
86. χ_5^2 = 31.13, p < 0.001
87. χ_3^2 = 25.2, p < 0.001
88. χ_1^2 = 1.67, p > 0.05
89. χ_5^2 = 16.798, p < 0.005
90. χ_5^2 = 1.34, p > 0.5 SEp = 99, Z = 1.26, p > 0.10
91. χ_1^2 = 2.64, p > 0.05
92. χ_1^2 = 11.77, p > 0.001
93. χ^2 = 0.812, p > 0.05
94. r = 0.9062, p < 0.001, b_{yx} = 3.01
    a = $\overline{Y} - b\overline{X}$ = 15.43
95. a. r = 0.9633, p < 0.01
    b. b_{yx} = 23.0768
    c. a = $\overline{Y} - b\overline{X}$ = 47.7539, Reg coeff equn $\overline{Y}$ = a + bx Y_c values are: 54.68, 56.82, 61.60, 66.22, 68.52
96. r = 0.9215, p < 0.001, b_{yx} = 0.4532, a = $\overline{Y} - b\overline{X}$ = –0.1560
98. Mid year population in 1973 = 353640 and in 1984 = 423837
101. Cholera morbidity rate = 0.11 and CFR = 25%
102. CBR = 30, IMR = 125, CDR = 14

103. SBR = 17.86, PMR = 40.51, NMR = 79.33, IMR = 123.96
105. Children per woman per year = 0.067
106. a. 86.94%   b. 75.244%   c. 58.539%
     d. 78.181%   e. 66.663%
108. $\chi^2 = 4.88$, DF = 1, $p < 0.05$
109. $\chi^2 = 14.51$, DF = 3, $p < 0.01$
110. $\chi^2 = 2.02$, DF = 3, $p < 0.05$

Chapter **18**

Computers in Medicine

Today we are in the midst of a computer revolution. Computers are becoming increasingly popular in every sphere of socio-economic or medical activity so also in medical establishments. It can safely be said that 21st century will be predominently ruled by computers. Still some of us are extremely sceptical about the idea of computers in medical practice.

Computer means a calculator but now it is a highly sophisticated tool not only for calculation but does many other jobs with tremendous speed. In the development process, the size of a computer has shrunk and the power has increased. A computer that was housed in a large hall in the late sixties is now seen in the form of **microcomputer** on a small table with the same power. A microcomputer indicates the tremendous computering (calculating) power of the computer. The computer's performance is to be compared to humans. A human percieves through the sensory organs, then the brain processes the input and there is an expression which may be vocal or a physical activity that depends on the training. Similarly, for a given input the Central Processing Unit (CPU) processes according to the programme to give the required output.

Application of computers in medical field may be outlined under four headings:
   I. Methods in biostatistics
  II. Community or public health care

III. Hospital establishments, nursing homes and clinics of academic general practitioners.
IV. Medical research

I. APPLICATION IN METHODS IN BIOSTATISTICS

This Chapter on "Computers in Medicine" is added to initiate the students and familiarise the research workers with the use of computers in biostatistics (as a science) described in Chapters 1 to 17.

The computers can be used and are used now in solving various problems in biostatistics for:
1. Collection, compilation, tabulation and diagrammatic presentation in the manner required for any size of data for completeness and accuracy.
2. Finding averages; coefficient of variation, standard deviation and standard error and percentiles, etc., of any size of data simultaneously.
3. The application of tests of significance such as 'Z','t' and χ^2, correlation and regression coefficients and other tests to find **probability** (P values), necessary in practice of medicine, public health and in research.
4. Construction of life tables to find longevity of life at birth and at any age such as on retirement for budgeting pension and other old age problems.
5. Probability of length of life after cancer without or after operation.
6. Chances of survival and period of survival after operation on heart such as bypass, closure of foramen ovale, repair and replacement of valves of heart, etc.
7. Successful chances of grafting of tissues such as cornea and parts of liver.
8. Survival after transplantation of heart, kidney and removal of brain tumour, etc.

World Registry of such surgical operations helps in above applications.

II. APPLICATION IN COMMUNITY OR PUBLIC HEALTH CARE AND MANAGEMENT

The management of public health care requires definition of objectives that are need based. To formulate policy, monitor implementation and finally to evaluate it, the prime need is **DATA**. For greater accuracy, quality data in large quantity must be precisely calibrated to assist in decision making. Computer based 'Health Information' systems have been described as a necessity by WHO.

Other familiar examples are our National Health Programmes:
1. Family Welfare rather family planning and MCH programmes
2. National Leprosy Prevention and Control Programme
3. National Tuberculosis Prevention and Control Programme
4. National Blindness Prevention and Control Programme
5. National Malaria Prevention and Control Programme
6. National Filaria Prevention and Control Programme
7. National Goitre Prevention and Control Programme
8. National Immunisation Programmes against few communicable diseases (EPI and UIP)
9. National Anaemia Prevention and Control Programme
10. National Diarrhoea Prevention and Control Programme
11. National Serious Respiratory Diseases Prevention and Control Programme
12. National Integrated Child Development Service (ICDS) Programme in which medical colleges, universities and health department of states are involved.

Computers have great capacity for storage of past and present data which helps in comparison of events in the past. A very familiar example is prevalence rate of tuberculosis which in 1958 and today in 1997 is same in spite of specific and effective drugs at our command and reduction in period of treatment from 3 years to 8 months allowing mobility and gradually normal work during treatment also. Perhaps it is due to socio-economic factors playing a prominent role.

Forecasting of epidemics like cholera, influenza, plague, viral hepatitis, malaria and many other communicable diseases was being done in the past by simple assumptions but now computers can help in that more precisely and effectively. That helps in modification of measures in prevention and control.

III. APPLICATIONS OF COMPUTERS IN HOSPITAL ESTABLISHMENTS, NURSING HOMES AND CLINICS OF ACADEMIC GENERAL PRACTITIONERS

To most doctors a computerised hospital means a place which generates bills, receipts, etc., on the printer rather than on a typewriter but it is not so. The advantages are to:
1. **Increase efficiency** and improve services to the patients for better professional care in computerised methods of diagnosis using cat scan angiography, biopsy before surgery and computer aided treatment as by laser in stone crushing in urinary system and electrical vaporisation in enlargement of prostate, etc., by better development of staff and resources and save time and money.
2. **Prevent losses** due to errors and pilferage and reduce overheads and salaries and tighter management of control to enhance image of establishment.
3. Computerise full range of drug reference systems by keeping check over stocks of drugs and their expiry date and standard recommendation of doses of different drugs.
4. **Save time** over non-clinical work like quick retrieval of records, spending less time on each revisit of the patient and quicker billing, etc.
5. Scheduling of appointments to reduce waiting periods for the patients, manage routine documents such as growth charts, follow-up of patients including progress, visits, immunisation, etc.
6. **Keeping academic references**—statistics and reference material for preparation and presentation of

papers and slides, thus keeping communications with others around the world.

Generally computers are used in the areas where humans are deficient like in fast calculations, storage and retrieving a large amount of information, searching for specific information and drawing for geographical distribution, colour manipulation, etc., Thus these are the areas where computers can help immensely.

Computers are very *versatile* in the areas such as **patients' records**, reports of investigations, inventory, billing and accounts.

Computerised medical records are very efficient, quick to locate, accurate space saving and long lasting. Computerised sheets are complete, systematic, consistent, legible, quick, less costwise and give a lot of information.

Problems with manual records are: loss of OPD data, repetitive data entry, reports of investigation including various reports like registration, laboratory reports, X-rays being lost and lack of standardisation, etc. This leads to a large amount of data being wasted.

Inventory It's extremely difficult to maintain inventory details manually especially with several thousand items like drugs, instruments, linen and furniture, etc. in the hospital. The result is that most hospitals are regularly throwing away large quantities of expired drugs and by the time, one is able to do the stock checking the status gets changed entirely.

Thus, computerised systems are helpful in checking, billing and accounts. Obviated long queus, totalling mistakes, wrong enteries and quarrelling of patients are the hallmarks of manual accounts. Bills of patients, staff and other things are computerised and such scenes are obviated.

So in all these fields, computers are of immense help and save manual labour, time as well as money.

Diagnosis

Computers can be used as an aid to assist practitioners to prescribe right medicine. Lots of information in terms of formulations available, symptoms, case histories, etc., can be stored and assessed by one press of a key. If the doctor himself is free he can make a probable diagnosis and give provisional treatment on the basis of probabilitis.

Today devices are available which can be connected to central computer at the Medical Centre remotely over telephone lines from a patient's home. Data such as blood pressure, weight, blood sugar levels, prenatal condition, foetal monitoring, smoking and drinking habit, etc., can be sent to one central computer using these specific devices. The same can be analysed and the patient guided over telephone line itself.

IV. COMPUTERS IN RESEARCH

Researchers in health care need to analyse data to understand and report significant observation for which large value of data must be analysed. Computers make this possible and in a short time. The development of application in programmes to support statistical analysis has made the task easier.

The computers still have not solved one problem, the right analysis for the data available.

This has lead to the development of **Statistical Expert System** to select the *appropriate statistical analysis* for the available data. Some of the more recent packages do incorporate these facilities.

To summarise the importance of understanding computers Milton H. Aronson is cited from the 1984 October issue of Bio Engineering. "A professional work in any field today faces a danger of obsolescence within his profession unless he understands the principles, practice and limitations of computers". The computer is not *"brain"* and *"smart"*. It is simply an *obedient servant*.

Appendices

Appendix I

Table of Unit Normal Distribution (UND) Single-tail

The table helps to find the percentage of area or observations in the normal distribution which lie beyond various standard deviation units ($Z = X-\bar{X}/\sigma$) from the mean. Z is 0 at the mean.

Z	0.00	0.01	0.02	0.03	0.04	0.05	0.06	0.07	0.08	0.09
0.0	0.5000	0.4960	0.4920	0.4880	0.4840	0.4801	0.4761	0.4721	0.4681	0.4641
0.1	0.4602	0.4562	0.4522	0.4483	0.4483	0.4404	0.4364	0.4325	0.4286	0.4247
0.2	0.4207	0.4168	0.4129	0.4090	0.4052	0.4013	0.3974	0.3936	0.3897	0.3859
0.3	0.3821	0.3783	0.3745	0.3707	0.3669	0.3632	0.3594	0.3557	0.3520	0.3483
0.4	0.3446	0.3409	0.3372	0.3336	0.3300	0.3264	0.3228	0.3192	0.3156	0.3121
0.5	0.3085	0.3050	0.3015	0.2981	0.2946	0.2912	0.2877	0.2843	0.2810	0.2776
0.6	0.2743	0.2709	0.2676	0.2643	0.2611	0.2578	0.2546	0.2514	0.2483	0.2451
0.7	0.2420	0.2389	0.2358	0.2327	0.2297	0.2266	0.2236	0.2206	0.2177	0.2148
0.8	0.2119	0.2090	0.2061	0.2033	0.2005	0.1977	0.1949	0.1922	0.1894	0.1867
0.9	0.1841	0.1814	0.1788	0.1762	0.1736	0.1711	0.1685	0.1660	0.1635	0.1611
1.0	0.1587	0.1562	0.1539	0.1515	0.1492	0.1469	0.1446	0.1423	0.1401	0.1379
1.1	0.1357	0.1335	0.1314	0.1292	0.1271	0.1251	0.1230	0.1210	0.1190	0.1170
1.2	0.1151	0.1131	0.1112	0.1093	0.1075	0.1056	0.1038	0.1020	0.1003	0.0985
1.3	0.0968	0.0951	0.0934	0.0918	0.0901	0.0885	0.0869	0.0853	0.0838	0.0823
1.4	0.0808	0.0793	0.0778	0.0764	0.0749	0.0735	0.0721	0.0708	0.0694	0.0681
1.5	0.0668	0.0655	0.0643	0.0630	0.0618	0.0606	0.0594	0.0582	0.0571	0.0559
1.6	0.0548	0.0537	0.0526	0.0516	0.0505	0.0495	0.0485	0.0475	0.0465	0.0455
1.7	0.0446	0.0436	0.0427	0.0418	0.0409	0.0401	0.0392	0.0384	0.0375	0.0367
1.8	0.0359	0.0351	0.0344	0.0336	0.0329	0.0322	0.0314	0.0307	0.0301	0.0294
1.9	0.0287	0.0281	0.0274	0.0268	0.0262	0.0256	0.0250	0.0244	0.0239	0.0233
2.0	0.0228	0.0222	0.2197	0.0212	0.0207	0.0202	0.0197	0.0192	0.0188	0.0183
2.1	0.0179	0.0174	0.0170	0.0166	0.0162	0.0158	0.0154	0.0150	0.0146	0.0143
2.2	0.0139	0.0136	0.0132	0.0129	0.0125	0.0022	0.0119	0.0116	0.0113	0.0110
2.3	0.0107	0.0104	0.0102	0.0099	0.0096	0.0094	0.0091	0.0089	0.0087	0.0084
2.4	0.0082	0.0080	0.0078	0.0075	0.0073	0.0071	0.0069	0.0068	0.0066	0.0064
2.5	0.0062	0.0060	0.0059	0.0057	0.0055	0.0054	0.0052	0.0051	0.0049	0 0048
2.6	0.0047	0.0045	0.0044	0.0043	0.0041	0.0040	0.0039	0.0038	0.0037	0.0036
2.7	0.0035	0.0034	0.0033	0.0032	0.0031	0.0030	0.0929	0.0028	0.0027	0.0026
2.8	3.0026	0.0025	0.0024	0.0023	0.0023	0.0022	0.0021	0.0021	0.0020	0.0019
2.9	0.0019	0.0018	0.0018	0.0017	0.0016	0.0016	0.0015	0.0015	0.0014	0.0014
3.0	0.0013	0.0013	0.0013	0.0012	0.0012	0.0012	0.0011	0.0011	0.0010	0.0010

Vertical column headed Z gives the Z values upto one decimal point such as 0.1, 0.2 etc., while horizontal values to the right of Z give the Z values upto two decimal points such as 0.00, 0.01, 0.02, etc.

To know the area or percentage of observations lying beyond the Z value of 1.96 find 1.9 in the Z column at the extreme left and then read across to the column headed 0.06. The area is 0.0250 hence the percentage of observations will be 2.5%.

Single tail means the area or observations lying to only one side of the curve mean beyond the deprived point where Z = 0. They lie—

to the left or beyond a particular observation when Z value is minus, or to the right or beyond a particular observation when the Z value is plus.

When Z is + 1.5, the area of the curve beyond or the percentage of observation **above** the value corresponding to Z is = −1.5, will be 0.0668 or 6.68%. Similarly when Z is = −1.5, the proportional of observations below the value corresponding to Z will be 6.68%.

Example

If the mean height $\bar{X}$ is 160 cm with SD of 5 cm, calculate Z value for an observation X 172.9 cm and interpret the result in the light of UND.

$$Z = \frac{172.9 - 160}{5} = 2.58$$

Table value for Z 2.58 = 0.0049. It means only 0.49% observations will exceed height 172.9 cm, if heights are normally distributed.

Conversely—

If the observed X is 147.1 cm.

$$\text{then } Z = \frac{147.1 - 160}{5} = -2.58$$

The table value for Z—2.58 is 0.0049. It means only 0.49% observations will be less than height 147.1 cm.

NB

'Z', 't' and 'r' tests are based on the assumption of normal distribution, They are called **parametric tests.**

An abridged table is constructed on the basis of UND which gives the probability (p) or relative frequency of occurrence of a Z value. P value helps to interpret the statistical significance of Z value.

P:	0.10	0.05	0.02	0.01
Z:	1.6	2.0	2.3	2.3

Non-parametric tests like χ^2, rank order correction coefficient and median tests do not require presumption of any specific distribution.

Appendix II

Table of 't'
Probability of Larger Value of 't'

DF	0.10	0.05	0.02	0.01	0.001
1	6.31	12.71	3.82	63.66	636.62
2	2.92	4.30	6.97	9.93	31.60
3	2.35	3.18	4.54	5.84	12.92
4	2.13	2.78	3.75	4.60	8.61
5	2.02	2.57	3.37	4.03	6.87
6	1.94	2.45	3.14	3.71	5.96
7	1.90	2.37	3.00	3.50	5.41
8	1.86	2.31	2.90	3.36	5.04
9	1.83	2.26	2.82	3.25	4.78
10	1.81	2.23	2.76	3.17	4.59
11	1.80	2.20	2.72	3.11	4.44
12	1.78	2.18	2.68	3.06	4.32
13	1.77	2.16	2.65	3.01	4.22
14	1.76	2.15	2.62	2.98	4.14
15	1.75	2.13	2.60	2.95	4.07
16	1.75	2.12	2.58	2.92	4.02
17	1.74	2.11	2.57	2.90	3.97
18	1.73	2.10	2.55	2.88	3.92
19	1.73	2.09	2.54	2.86	3.88
20	1.73	2.09	2.53	2.85	3.85
21	1.72	2.08	2.52	2.83	3.82
22	1.72	2.07	2.51	2.82	3.79
23	1.71	2.07	2.50	2.81	3.77
24	1.71	2.06	2.49	2.80	3.75
25	1.71	2.06	2.49	2.79	3.73
26	1.71	2.06	2.48	2.78	3.71
27	1.70	2.05	2.47	2.77	3.69
28	1.70	2.05	2.47	2.76	3.67
29	1.70	2.05	2.46	2.76	3.66
30	1.69	2.04	2.46	2.75	3.65
40	1.68	2.02	2.42	2.70	3.55
60	1.67	2.00	2.39	2.66	3.46
120	1.66	1.98	2.36	2.62	3.37
∞	1.65	1.96	2.33	2.58	3.29

The table gives the probability of observing the highest 't' value by chance at particular degrees of freedom. The probability of observing value of 't' greater than 2.95 at 15 degrees of freedom is 0.01 or 1%.

Appendix III

Variance Ratio
5 Per Cent, Points of e^{22}

n_1/n_2	1	2	3	4	5	6	8	12	24	∞
1	161.40	199.5	215.7	224.6	230.2	234.0	238.9	243.9	249.0	254.3
2	18.51	19.00	19.16	19.25	19.30	19.33	19.37	19.41	19.45	19.50
3	10.13	9.55	9.28	9.12	9.01	8.94	8.84	8.74	8.64	8.53
4	7.71	6.94	6.59	6.39	6.26	6.16	6.04	5.91	5.77	5.63
5	6.61	5.79	5.41	5.19	5.05	4.95	4.82	4.68	4.53	4.36
6	5.99	5.14	4.76	4.53	4.39	4.28	4.15	4.00	3.84	3.67
7	5.59	4.74	4.35	4.12	3.97	3.87	3.73	3.57	3.41	3.23
8	5.32	4.46	4.07	3.84	3.69	3.58	3.44	3.28	3.12	2.93
9	5.12	4.26	3.86	3.63	3.48	3.37	3.23	3.07	2.90	2.71
10	4.96	4.10	3.71	3.48	3.33	3.22	3.07	2.91	2.74	2.54
11	4.84	3.98	3.59	3.36	3.20	3.09	2.95	2.79	2.61	2.40
12	4.75	3.88	3.49	3.26	3.11	3.00	2.85	2.69	2.50	3.30
13	4.67	3.80	3.41	3.18	3.02	2.92	2.77	2.60	2.42	2.21
14	4.60	3.74	3.34	3.11	2.96	2.85	2.70	2.53	2.35	2.13
15	4.54	3.68	3.29	3.06	2.90	2.79	2.64	2.48	2.29	2.07
16	4.49	3.63	3.24	3.01	2.85	2.74	2.59	2.42	2.24	2.01
17	4.45	3.59	3.20	2.96	2.81	2.70	2.55	2.38	2.19	1.96
18	4.41	3.55	3.16	2.93	2.77	2.66	2.51	2.34	2.15	1.92
19	4.38	3.52	3.13	2.90	2.74	2.63	2.48	2.31	2.11	1.88
20	4.35	3.49	3.10	2.87	2.71	2.60	2.45	2.28	2.08	1.84
21	4.32	3.47	3.07	2.84	2.68	2.57	2.42	2.25	2.05	1.81
22	4.30	3.44	3.05	2.82	2.66	2.55	2.40	2.23	2.03	1.78
23	4.28	3.42	3.03	2.80	2.64	2.53	2.38	2.20	2.00	1.76
24	4.26	3.40	3.01	2.78	2.62	2.51	2.36	2.18	1.98	1.73
25	4.24	3.38	2.99	2.76	2.60	2.49	2.34	2.16	1.96	1.71
26	4.22	3.37	2.98	2.74	2.59	2.47	2.32	2.15	1.95	1.69
27	4.21	3.35	2.96	6.73	2.57	2.46	2.30	2.13	1.93	1.67
28	4.20	3.34	2.95	2.71	2.56	2.44	2.29	2.12	1.91	1.65
29	4.18	3.33	2.93	2.70	2.54	2.43	2.28	2.10	1.90	1.64
30	4.17	3.32	2.92	2.69	2.53	2.42	2.27	2.09	1.89	1.62
40	4.08	3.23	2.84	2.61	2.45	2.34	2.18	2.00	1.79	1.51
60	4.00	3.15	2.76	2.52	2.37	2.25	2.10	1.92	1.70	1.39
120	3.92	3.07	2.68	2.45	2.29	2.17	2.02	2.83	1.61	1.25
∞	3.84	2.99	2.60	2.37	2.21	2.10	1.94	1.75	1.52	0.00

Lower 5 per cent points are found by interchange of n_1 and n_2. i.e. n_1 must always correspond with the greater mean square.

Variance Ratio—*continued*
1 Per Cent. Points of e^3z

n_1/n_2	1	2	3	4	5	6	8	12	24	∞
1	4052	4999	5403	5625	5764	5859	5982	6106	6234	6366
2	98.50	99.00	99.17	99.25	99.30	99.33	99.37	99.42	99.46	99.50
3	34.12	30.82	29.46	28.71	28.24	27.91	27.49	27.05	26.60	26.12
4	21.20	18.00	16.69	15.98	15.52	15.21	14.80	14.37	13.93	13.46
5	16.26	13.27	12.06	11.39	10.97	10.67	10.29	9.89	9.47	9.02
6	13.74	10.92	9.78	9.15	8.75	8.47	8.10	7.72	7.31	6.88
7	12.25	9.55	8.45	7.85	7.46	7.19	6.84	6.47	6.07	5.65
8	11.26	8.65	7.59	7.01	6.63	6.37	6.03	5.67	5.28	4.86
9	10.56	8.02	6.99	6.42	6.06	5.80	5.47	5.11	4.73	4.31
10	10.04	7.56	6.55	5.99	5.64	5.39	5.06	4.71	4.33	3.91
11	9.65	7.20	6.22	5.67	5.32	5.07	4.74	4.40	4.02	3.60
12	9.33	6.93	5.95	5.41	5.06	4.82	4.50	4.16	3.78	3.36
13	9.07	6.70	5.74	5.20	4.86	4.62	4.30	3.96	3.59	3.16
14	8.86	6.51	5.56	5.03	4.69	4.46	4.14	3.80	3.43	3.00
15	8.68	6.36	5.42	4.89	4.56	4.32	4.00	3.67	3.29	2.87
16	8.53	6.23	5.29	4.77	4.44	4.20	3.89	3.55	3.18	2.75
17	8.40	6.11	5.18	4.67	4.34	4.10	3.79	3.45	3.08	2.65
18	8.28	6.01	5.09	4.58	4.25	4.01	3.71	3.37	3.00	2.57
19	8.18	5.93	5.01	4.50	4.17	3.94	3.63	3.30	2.92	2.49
20	8.10	5.85	4.94	4.43	4.10	3.87	3.56	2.23	2.86	2.42
21	8.02	5.78	4.87	4.37	4.04	3.81	3.51	3.17	2.80	2.36
22	7.94	5.72	4.82	4.31	3.99	3.76	3.45	3.12	2.75	2.31
23	7.88	5.66	4.76	4.26	3.94	3.71	3.41	3.07	2.70	2.26
24	7.82	5.61	4.72	4.22	3.90	3.67	3.36	3.03	2.66	2.21
25	7.77	5.57	4.68	4.18	3.86	3.63	3.32	2.99	2.62	2.17
26	7.72	5.53	4.64	4.14	3.82	3.59	3.29	2.96	2.58	2.13
27	7.68	5.49	4.60	4.11	3.78	3.56	3.26	2.93	2.55	2.10
28	7.64	5.45	4.57	4.07	3.75	3.53	3.23	2.90	2.52	2.06
29	7.60	5.42	4.54	4.04	3.73	3.50	3.20	2.87	2.49	2.03
30	7.56	5.39*	4.31	4.02	3.70	3.17	3.47	2.84	2.47	2.01
40	7.31	5.18	4.31	3.83	3.51	3.29	2.99	2.66	2.29	1.80
60	7.08	4.98	4.13	3.65	3.34	3.12	2.82	2.50	2.12	1.60
120	6.85	4.79	3.95	3.48	3.17	2.96	2.66	2.34	1.95	1.38
∞	6.64	4.60	3.78	3.32	3.02	2.80	2.51	2.18	1.79	1.00

Lower 1 per cent. points are found by interchange of n_1 and n_2, i.e., n_1 must always correspond with the greater mean square.

Appendix IV

Table of χ^2
Probability (P)

DF	0.50	0.10	0.05	0.02	0.01	0.001
1	0.46	2.71	3.84	5.41	6.64	10.83
2	1.39	4.61	5.99	7.82	9.21	13.82
3	2.37	6.25	7.82	9.84	11.34	16.27
4	3.36	7.78	9.49	11.67	13.28	18.47
5	4.35	9.24	11.07	13.39	15.09	20.52
6	5.35	10.65	12.59	15.03	16.81	22.46
7	6.35	12.02	14.07	16.62	18.48	24.32
8	7.34	13.36	15.51	18.17	20.09	16.13
9	8.34	14.68	16.92	19.68	21.67	27.88
10	9.34	15.89	18.31	21.16	23.21	29.59
11	10.34	17.28	19.68	22.62	24.73	31.26
12	11.34	18.55	21.03	24.05	26.22	32.91
13	12.34	19.81	22.36	25.47	27.69	34.53
14	13.34	21.06	23.69	26.87	29.14	36.12
15	14.34	22.31	24.99	28.26	30.58	37.70
16	15.34	23.54	26.30	29.63	32.00	39.25
17	16.34	24.77	27.59	30.99	33.41	40.79
18	17.34	25.99	28.87	32.35	34.81	42.31
19	18.34	27.20	30.14	33.69	36.19	43.82
20	19.34	28.41	31.41	35.02	37.57	45.32
21	20.34	29.62	32.67	36.34	38.93	46.80
22	21.34	30.81	33.92	37.66	40.29	48.27
23	22.34	32.01	35.17	38.97	41.64	49.73
24	23.34	33.20	36.42	40.27	42.98	51.18
25	24.34	34.38	37.65	41.57	44.31	52.62
26	25.34	35.56	38.89	42.86	45.64	54.05
27	26.34	36.74	40.11	44.14	46.96	55.48
28	27.34	37.92	41.34	45.42	48.28	56.89
29	28.34	39.09	42.56	46.69	49.59	58.30
30	29.34	40.26	43.77	47.96	50.89	59.70

The table gives the highest values of χ^2, at particular degrees of freedom, corresponding to probability P of occurrence by chance in nature, e.g., at 10 degrees of freedom χ^2 value larger than 18.31 will occur less than 5 times in 100 (P<0.05) and is interpreted as significance at 5% level.

Appendix V

The Correlation Coefficient Table
Probability (P)

DF	0.10	0.05	0.02	0.01	0.001
1	0.9877	0.9969	0.9995	0.99988	0.99999
2	0.9000	0.9500	0.9800	0.9900	0.9990
3	0.805	0.878	0.9343	0.9587	0.9912
4	0.729	0.811	0.882	0.9172	0.9741
5	0.669	0.755	0.833	0.875	0.9507
6	0.621	9.707	0.789	0.834	0.925
7	0.582	0.666	0.750	0.798	0.898
8	0.549	0.632	0.716	0.765	0.872
9	0.521	0.602	0.685	0.735	0.847
10	0.497	0.576	0.658	0.708	0.823
11	0.476	0.553	0.634	0.684	0.801
12	0.457	0.532	0.612	0.661	0.780
13	0.441	0.514	0.592	0.641	0.760
14	0.426	0.497	0.574	0.623	0.742
15	0.412	0.482	0.558	0.606	0.725
16	0.400	0.468	0.543	0.590	0.708
17	0.389	0.456	0.529	9.575	0.693
18	0.378	0.444	0.516	0.561	0.679
19	0.369	0.433	0.503	0.549	0.665
20	0.360	0.423	0.492	0.537	0.652
25	0.323	0.381	0.445	0.487	0.597
30	0.296	0.349	9.409	0.449	0.554
35	0.275	0.325	0.381	0.418	0.519
40	0.257	0.304	0.358	0.393	0.490
45	0.243	0.288	0.338	0.372	0.465
50	9.231	0.273	0.322	0.354	0.443
60	0.211	0.250	0.295	0.325	0.408
70	0.195	0.232	0.274	0.302	0.380
80	0.183	0.217	0.257	0.283	0.357
90	0.173	0.205	0.242	0.267	0.338
100	0.164	0.195	0.230	0.254	0.321

The table gives percentage points for the distribution of the estimated correlation coefficient (r) when true value (r) is zero. Thus, when there are 10 degrees of freedom (i.e., in a sample of 12) the probability of observing, r greater in absolute value than 0.576, by chance is 0.05 or 5 per cent.

In this case DF = Degrees of freedom = n–2.

Appendix VI
Random Numbers

03	47	43	73	86	36	96	47	36	61	46	98	63	71	62	33	26	16	80	45
97	74	24	67	62	42	81	14	57	20	42	53	32	37	32	27	07	36	07	51
16	76	63	27	66	56	50	26	71	07	31	90	79	78	53	13	55	38	58	59
12	56	85	99	26	96	96	68	27	31	05	03	72	93	15	57	12	10	14	21
55	59	56	35	64	38	54	82	46	22	31	62	43	09	90	06	18	44	32	53
16	22	77	94	39	49	54	43	54	82	17	37	93	23	78	87	35	20	96	43
84	42	17	53	31	57	24	55	06	88	77	04	74	47	67	21	76	33	50	25
63	01	63	78	59	16	95	55	67	19	98	10	50	71	75	12	86	73	58	07
33	21	12	34	29	78	64	56	7	82	52	42	07	44	38	15	51	00	13	42
57	60	86	32	44	09	47	27	96	54	49	17	46	09	62	90	52	84	77	27
18	18	07	92	46	44	17	16	58	09	79	83	86	19	62	06	76	50	03	10
26	62	38	97	75	81	16	07	44	99	83	11	46	32	24	20	14	85	88	45
23	42	40	64	74	82	97	77	77	81	07	45	32	14	08	32	98	94	07	72
62	36	28	19	95	50	92	26	11	97	00	56	76	31	38	80	22	02	53	53
37	85	94	35	12		39	50	08	30	42	34	07	96	88	54	42	06	87	98
70	29	17		53	13	33	20	38	26	13	89		03	74	17	76	37	13	04
56	62	18	37	35	96	83	50	87	75	97	12	25	93	47	70	33	24	03	54
99	49	57	22	77	88	42	95	45	72	16	64	36	16	00	04	43	18	66	79
16	08	15	04	72	33	27	14	34	09	45	59	34	68	49	12	72	07	34	45
31	16	93	32	43	50	27	89	87	19	20	15	37	00	49	52	85	66	90	11

contd.

68	34	30	13	70	55	74	30	77	40	44	22	78	84	26	04	33	46	09	52
74	57	25	65	76	59	29	97	68	60	71	91	38	67	54	13	58	18	24	76
27	42	37	86	53	48	55	90	65	72	96	57	69	36	10	96	46	92	42	45
00	39	68	29	61	66	37	32	20	30	77	84	57	03	29	10	45	65	04	26
29	94	98	94	24	68	49	69	10	82	53	75	91	93	30	34	25	20	57	27
16	90	82	66	59	83	62	64	11	12	67	19	00	71	74	60	47	21	29	68
11	27	94	75	06	06	09	19	74	66	02	94	37	34	02	76	70	90	30	86
35	24	10	16	20	33	32	51	26	38	79	78	45	04	91	16	92	53	56	16
38	23	16	86	38	42	38	97	01	50	87	75	66	81	41	40	01	74	91	62
31	96	25	91	47	96	44	33	49	13	34	86	82	53	91	00	52	43	48	85
66	67	40	67	14	64	05	71	95	86	11	05	65	09	68	76	83	20	37	90
14	90	84	45	11	75	73	88	05	90	52	27	41	14	86	26	98	12	22	08
68	05	51	18	00	33	96	02	75	19	07	60	62	93	55	59	33	82	43	90
20	46	78	73	90	97	51	40	14	02	04	02	33	31	08	39	54	16	49	36
64	19	58	97	79	15	06	15	93	20	01	90	10	75	06	40	78	78	89	62
05	26	93	70	60	22	35	85	15	13	92	03	51	59	77	59	56	78	06	83
07	97	10	88	23	09	98	42	99	64	61	71	62	99	15	06	51	29	16	93
68	71	86	85	85	54	87	66	47	54	73	32	08	11	12	44	95	92	63	16
26	99	61	65	53	58	37	78	80	70	42	10	50	67	42	32	17	55	85	74
14	65	52	68	75	87	59	36	22	41	26	78	63	06	55	13	08	26	01	5

Appendix VII
International Classification of Disease

In the last edition of *Methods of Biostatistics* 1989 which was published by DGHS for India in 1979. Three diseases code ranging from 001–999 were given in disease categories.

This classification was revised by World Health Assembly (WHA) in 1990, 1991 which was also attended by Dy. Director General, Dr. S.S. Srivastav, Central Statistical Organisation on behalf of India.

This classification was amended by DGHS subsequently. It is in three volumes hence not published in the revised edition but the reader may refer for the same to Central Statistical Organisation, Patel Bhavan, New Delhi-1.

Moreover, the International Classification is not there in the new edition as it does not necessarily form part of methods in biostatistics or even vital statistics.

Bibliography

Armitage P.	Statistical Method in Medical Research
Aviva Petrie	Lecture Notes on Medical Statistics
Bailey, N.T.J.	Statistical Methods in Biology
Bernstein, L. and Weatherall. M.	Statistics for Medical and other Biological Students
Bisht, D.B.	Basic Principles of Medical Research
Bradford Hill, A.	Principles of Medical Statistics
Burn, J.H.	Biological Standardisation, Ch. 3
Fisher, R.A.	Statistical Methods for Research Workers
Fisher, R.A.	Design of Experiments for Research Workers (Oliver and Boyd)
Fisher, R.A. and Yates, F.	Statistical Tables (Oliver and Boyd)
Geoffrey J. Bourke and James, M.C.	Interpretation and Uses of Medical Statistics
Goulden, C.H.	Methods of Statistical Analysis (Wiley)

Govt. of India, Ministry of H. & F.W., New Delhi, Year Book 1985–86, F.W.P. in India.

Hubert, M. Blalock, J.R.	Social Statistics
Huldah Bancroft	Introduction to Biostatistics
Leonard A. Coldstone	E.L.B.S.—1985, Understanding Medical Statistics.
Lilian Cohn	Statistical Methods for Social Scientists
Lynn Smith, T.	Demography, Principles and Methods
Mainland, D.	Elementary Medical Statistics
Mase Well, A.E.	Analysing Qualitative data
Mathen, K.K.	A Guide to Health Statistics
Rao. N.S.N.	Elements of Health Statistics

Robert, C. Duncen Rebecca G. Knapp *et al*	Introductory Biostatistics for the Health Sciences
Satya Swaroop	Introduction to Health Statistics
Snedecor, G.W.	Statistical Methods

United Nations N. York Handbook of Vital Statistics Methods

Winifred, M. Castle   Statistics in Small Doses, 1979.

WHO, Geneva, Edited by S.K. Lawanga & Cho-Yook Tye, Teaching Health Statistics. Twenty Lessons and Seminar Outlines
WHO Regional Office for S.E.A., New Delhi 1986
Bulletin of Regional Health Information, 1984–85.

Index

A

Association 169
Averages 35

B

Bar diagram 30
Binomial classification 157
Binomial frequency
 distribution 159
Binomial probability
 distribution 108
Birth rates 251
Bivariate frequency distribution
 table 161, 187, 329

C

Case fatality rate 268
Child-women ratio 247
Chi-Square table 331
Chi-Square test 168
 as a test of association 169
 as a test of difference in
  proportions 168
 as a test of goodness of fit 171
 restriction in 174
Coefficient of variation 74
Cohort 208
Confidence interval 118
Confidence limits 118
Correlation 187
 calculation of 190
 coefficient 190
 table 332
Cross sectional studies 28
Cumulative frequency diagram 28
Cumulative frequency table 27

D

Data 11

presentation of 14
qualitative 11
quantitative 12
Death certificate 225
Death rates 260
 age-sex specific 262
 cause specific 267
 crude 260
 specific 262
Degrees of freedom 141
Demography 214
Distribution 16
 asymmetrial 85
 binomial 158
 normal 79
 sampling 115
Double blind trial 209

E

Error Type I and Type II 122
Expectation of Life 262

F

Foetal death 217
 ratio 252
Fertility rate 251
Frequency curve 24
Frequency distribution table 15
Frequency polygon 23
Frequency table 15

G

Group interval 19

H

Histogram 20

I

Illegitimate fertility rate 255

Immaturity 223
Incidence rate 256
Infant mortality rate 263
Instrumental error 62
Inter-quartile range 65

L

Level of significance 119
Life table 275
    modified 286
Line chart 25
Live birth 222
Longevity of life 222, 266, 276, 279

M

Map diagram 33
Master table 211
Marriage rate 255
Maternal mortality rate 266
Mean 36
    of grouped series 41
    of ungrouped series 38
Mean deviation 66
Measures of central tendency 36
Measures of dispersion 63
Measures of health 273
Measures of location 48
Median 37
    of grouped series 48
    of ungrouped series 47
Mid-year population 236
Mode 38
Morbidity rates 256

N

Neonatal mortality rate 264
Normal curve 81
Normal deviate 82
Normal distribution 77
Notations 8
Null hypothesis 121

O

Observer error 61

Ogive 27

P

Parameter 8
Percentile 48
    uses of 53
Perinatal mortality rate 264
Pictogram 33
Pie diagram 32
Poison distribution 14
Population 236
    age-sex distribution 246
    arithmetical progression 239
    census 215
    density 246
    geometrical progression 240
    natural increase 239
    pyramid 248
Post-neonatal mortality rate 264
Pregnancy rate 269
Pregnancy prevalence rate 270
Prevalence rate 257
Probability 104
    addition law 105
    multiplication law 107
    from normal distribution 112
Proportional mortality
    indicator 272
Proportional mortality rate 272
Proportions 156

Q

Quartiles 49

R

Random sample 96
Random sampling numbers 333, 334
Range 63
Rates, crude, specific 249
Records of vital events 216
Regression coefficient 195
    calculation of 197
    line 198
Reproduction rate 253

S

Sample 8
  bias 94
  precision 89
  size 90
  representative 89
Sampling 88
  cluster 100
  multiphase 101
  multistage 100
  simple random 96
  stratified 99
  systematic 98
Sampling defects 62
Sampling distribution 115
Sampling techniques 96
Scatter diagram 29
Sector diagram 32
Selection 88
Semi-interquartile range 65
Significance 117
Spot map 32
Standard deviation 66
  of regression 199
Standard error 131
  of diff. between two means 135
  of diff. between two
    proportions 163
  of mean 131
  of proportion 158
Statistic 8
Statistical methods 2
Still-birth 223
  rate 264
Symbols 9

T

Tests of significance 125
't' table 328

't' test 141
  paired 147
  unpaired 142

U

Unit normal distribution,
  table 326

V

Variance ratio test 151
Variance 69
Variability 58
  biological 58
  class or group 60
  experimental 61
  individual 60
  measures of 63
  periodical 60
  real 60
  sampling 60

W

Weight card 54
Weighted mean 41

Y

Yates correction 174

Z

Z test 82, 125, 126
  one-tailed 129
  two-tailed 128